Medicinal Plants and Cancer Chemoprevention

Cancer is the second leading cause of death globally. *Medicinal Plants and Cancer Chemoprevention* provides information on the use of various herbal plants used as anti-cancer agents. It discusses the traditional system of medicine and focuses on plant-derived compounds for cancer therapy with integrated approaches. It presents information on various medicinal plants that covers background and history, ethnomedical considerations, morphology, phytochemistry, and pharmacological properties. The book also includes scientific investigation on medicinal plants in managing cancer, reported mechanisms of action as anti-cancer activity, and the toxicological aspects of certain plants.

Key Features:

- Details information on plant-derived compounds for cancer therapy
- Features information on methods of extraction and isolation of various phytoconstituents responsible for anti-cancer activity
- Discusses herbal formulations and alternative approaches used for the management and treatment of cancer
- Demonstrates the importance of alternative approaches including yoga, acupuncture, and dietary supplements to be effective in the management of cancer

This book is helpful to botanists, researchers, and practitioners in alternative and complementary medicine, and the herbal medicine research community.

Medicinal Plants and Cancer Chemoprevention

Edited by
Dr. Sapna Malviya and Dr. Neelesh Malviya

Assisted by
Ankur Joshi, Varsha Johariya, and Rajiv Saxena

CRC Press is an imprint of the
Taylor & Francis Group, an **informa** business

First edition published 2024
by CRC Press
6000 Broken Sound Parkway NW, Suite 300, Boca Raton, FL 33487-2742

and by CRC Press
4 Park Square, Milton Park, Abingdon, Oxon, OX14 4RN

CRC Press is an imprint of Taylor & Francis Group, LLC

ISBN: 978-1-032-17076-3 (hbk)
ISBN: 978-1-032-17075-6 (pbk)
ISBN: 978-1-003-25171-2 (ebk)

DOI: 10.1201/9781003251712

Typeset in Times
by KnowledgeWorks Global Ltd.

This book is gratefully dedicated to my loving and caring mother, Nirmala Motwani (Educationalist).

Contents

About the Book

This book covers information about causes, symptoms, risk factors, and molecular biology in the prevention, management, and treatment of cancer. It contains all the information about the traditional system of medicines and plant-derived compounds for cancer therapy with integrated approaches. The book contains all major plant constituents and scientific validation of traditional medicine with its mechanism of action against cancer. The book contains information about many medicinal plants, including their background, ethnomedical considerations, morphology and pharmacognostical character, phytochemistry, pharmacological properties, scientific investigation for the management of cancer, reported mechanism of action as anti-cancer activity, and toxicological aspects. Indeed, the book covers various methods of extraction and isolation of the phytoconstituents responsible for anti-cancer activity. The herbal industry is continuously in demand looking at this parameter. Various herbal formulations are mentioned in the book which are used for the treatment and management of cancer. In addition, the book contains alternative approaches for the treatment and prevention of cancer. So the book comprises all information about cancer with the remedial solutions for its treatment. Hopefully, the reader will gather all information desired.

Sapna Malviya
Neelesh Malviya

Preface

Medicinal Plants and Cancer Chemoprevention is the result of abundant hard work of the authors. They have given extensive details about medicinal plants for treatment of cancer. The book highlights the traditional system of medicines with information about plant-derived compounds for cancer therapy.

The authors express their deep gratitude to their mother, father, grandfathers, and son for motivating them to write a book for undergraduate, postgraduate, and research students.

Undoubtedly, authors are indebted to all who have supported in giving the present shape to the book. Suggestions and criticism will always be solicited by the authors to further improve the quality of the book in a real sense.

Sapna Malviya
Neelesh Malviya

About the Editors

Prof. Dr. Sapna Malviya, M. Pharm. (Hons.), MBA (HR), Ph.D. (Pharm. Sci.), Professor & Head of Institute, Modern Institute of Pharmaceutical Sciences, Indore, India

Prof. Dr. Sapna Malviya attained grants from the All India Council for Technical Education (AICTE), Madhya Pradesh Council of Science & Technology (MPCST), Bhopal, Atomic Energy Regulatory, Mumbai, and Entrepreneurship Development Institute of India, Ahmedabad, India, in the field of pharmaceutical sciences. She is Innovation Ambassador, MOE Innovation Cell, Government of India, and Institute Innovation Council, Ministry of Education Initiative, Chief Mentor and Advisor at MSME-sponsored Modern Incubator. She has received her pharmaceutical education from B. R. Nahata College of Pharmacy, Mandsaur (Madhya Pradesh). She is Qualified ZED Consultant to Pharmaceutical Industries.

An academician par excellence, she has superlative research expertise in Pharmaceutical Sciences. She has B. Pharm and M. Pharm degrees. She also has Postgraduate Degree in Management. Later, she completed her Doctorate in Pharmaceutical Sciences. She is presently working as Head of Institute, Pharmacy, Head of R & D Cell, Mentor at MSME-sponsored Modern Incubator, and other portfolios in Modern Institute of Pharmaceutical Sciences.

She has the distinguished honour of guiding many undergraduate and postgraduate students in their project and research work. She is Chief Coordinator and Convenor of many National Seminars, Co-Chairman of Inaugural Committee of Indian Pharmaceutical Association, MP State Branch Indore, Speaker in Faculty Development Programme, and Executive Development Programme Coordinator of various seminars and lectures.

Her scientific endeavours include Best Jury Award in 7th Academic Brilliance Awards-19 organized by EET CRS Research wing for Excellence in Professional Education & Industry, Delhi, in 2019; Best Teacher of Pharmacognosy, awarded by APTI MP State Branch, Bhopal, Excellence Award in MPCST-sponsored National Seminar, as well as Excellent Academician & Researcher Award of the Year in Central India in the recognition of excellence in Academic practices and research excellence in the field of pharmaceutical science in international conference. She has received Distinguished Professor Award in DST-sponsored conference and IPA, MP State Award ceremony 2021.

Dr. Malviya's untainted achievements are recognized by her publications in various revered peer-reviewed national and international journals which are more than a hundred in number. She has authored many textbooks and chapters for well-renowned publishers like CBS, presented research papers, written articles and columns in newspapers on pharmacy education, and guided postgraduate and graduate research projects to date. She is Honorary Life Member of IPA and APTI. She is

State President of the Women's Forum of SPSSR. Dr. Malviya has organized many conferences/workshops/seminars/FDPs/EDP/EAC. She has given research consultancy and training programmes in pharma institutions as well as industries. She has received a patent on poly herbal formulation for treatment of Alzheimer's and eight design patents for various pharmaceutical instruments. She has actively been involved in various preclinical studies on diabetes mellitus, assessment of various pharmacological activities of plant drugs, and phytopharmacological evaluation of plant drugs for antidiabetic, antiobesity, and various other health disorders.

Dr. Malviya has special acumen in planning, organizing, directing, and executing events as she has coordinated various events of Indian Pharmaceutical Association (IPA) and Association of Pharmaceuticals Teachers of India (APTI). She has anchored various symposiums and forums.

Prof. Dr. Neelesh Malviya, M. Pharm. (Hons.), MBA (HR), Ph.D. (Pharm. Sci.), Professor & Principal, Smriti College of Pharmaceutical Education, Indore, India

Prof. Dr. Neelesh Malviya attained a research project from the Department of Science & Technology, New Delhi (DST), All India Council for Technical Education (AICTE), Madhya Pradesh Council of Science & Technology, Bhopal (MPCST), due to his research aptitude in the field of pharmaceutical sciences. He earned his entire pharmaceutical education from B.R. Nahata College of Pharmacy, Mandsaur (Madhya Pradesh). His academic excellence includes various honours such as university medal in M. Pharm course by RGPV University in all branches among all affiliated pharmacy institutions of Madhya Pradesh and being institution topper in his Bachelor's and Master's degrees.

His scientific accomplishments include the Distinguished Professor Award in DST-NSTMIS Stakeholder Workshop and Best Director Award in 7th Academic Brilliance Awards-19 organized by EET CRS Research wing for Excellence in Professional Education & Industry, Delhi, in 2019, Innovative Academician & Researcher of the Year in Central India in 2018 by Innovative Pharmacist Group, Outstanding Researcher Award in 2018, Young Scientist Award in 2017 by Centre for Education Growth and Research, New Delhi, and best paper presentation award in various conferences.

In addition, Dr. Malviya is acknowledged with a Certificate of Honour in mentoring students to achieve highest results in entire RGPV-affiliated pharmacy institutions of Madhya Pradesh by Srizan in the academic year 2016–2017 and 2017–2018.

The major areas of his research are phytochemistry and herbal drug development. He has associated as external expert with reputed pharma companies for providing expertise for formulation development and optimization. He has published one patent on phytopharmacological screening of herbal medicines and has filed eight patent designs on laboratory equipment designs and signed MOU with Tata Strive, Neo Drugs—manufacturer of herbal products, Saify Healthcare & Medi-Devices India Private Limited, Indo Soviet Friendship College of Pharmacy (ISFCP), Moga, and Atal Incubation Centre, Indore.

Dr. Malviya's achievements are recognized by his publications in various reputed peer-reviewed national and international journals which are more than a hundred in number. He has authored ten textbooks and eight chapters, presented more than a hundred research papers, written articles and columns in newspapers on pharmacy education, and guided 25 postgraduate and 23 graduates research projects to date. Presently supervising eight PhD and six postgraduate research projects, he is Honorary Life Member of IPA, APTI, Indian Society of Pharmacognosy, ISTE, IPGA, CEGR, and Society of Ethnopharmacology.

He has organized more than 50 conferences/workshops/seminars/FDPS. He has given more than ten research consultancy and training programmes in pharma institutions as well as industries.

Contributors

Manisha Dhere
Assistant Professor
Department of Pharmaceutics
Smriti College of Pharmaceutical Education
Indore, India

Anindya Goswami
Associate Professor
Department of Pharmaceutics
Smriti College of Pharmaceutical Education
Indore, India

Ruchi Gupta
Associate Professor
Department of Pharmacognosy
Smriti College of Pharmaceutical Education
Indore, India

Varsha Johariya
Assistant Professor
Department of Medicinal Chemistry
Modern Institute of Pharmaceutical Sciences
Indore, India

Ankur Joshi
Associate Professor
Department of Pharmacology
Modern Institute of Pharmaceutical Sciences
Indore, India

Ashok Koshta
Associate Professor
Department of Pharmaceutics
Modern Institute of Pharmaceutical Sciences
Indore, India

Rajiv Saxena
Associate Professor
Department of Biotechnology
Smriti College of Pharmaceutical Education
Indore, India

Anamika Singh
Associate Professor
Department of Pharmaceutics
Modern Institute of Pharmaceutical Sciences
Indore, India

Priyanka Soni
Associate Professor
Department of Pharmaceutical Chemistry
Chameli Devi Institute of Pharmacy
Indore, India

1 Introduction to Cancer

Varsha Johariya, Ankur Joshi, Neelesh Malviya, and Sapna Malviya

CONTENTS

1.1 INTRODUCTION

Cancer refers to the abnormal growth of cell tissue which emerges from the change of typical cells into tumour cells in a multistage procedure that by and large advances from pre-destructive injury to a malignant tumour. Cancer is usually split as benign or malignant. A benign tumour is localized, develops slowly, and usually does not lead to death. Malignant or cancerous tumours develop more rapidly. They are not localized and are sometimes fatal. Cancer cells can break away from the original mass of cells, migrate through the blood and lymph systems, and lodge in other organs, where they can begin their uncontrolled growth cycle. Metastatic expansion, or metastasis, is the process of cancer cells leaving one zone and spreading to another area of the body.

DOI: 10.1201/9781003251712-1

Cancer is a genetic disorder that is triggered by gene changes that regulate the way our cells function, and particularly they expand and divide. Genetic changes can be inherited from cancer, and they can be inherited from our parents and cause cancer. They can also happen during a person's lifetime as a result of an error that occurs as cells divide or as a result of DNA damage caused by certain environmental exposures. Substances, such as the chemicals in cigarette smoke, and radiation, such as ultraviolet rays from the sun, are examples of cancer-causing environmental exposures. Each person's cancer has a unique combination of genetic changes.

Cancer cells may arise from normal cells. Until cancer cells mature in body tissues, cells undergo abnormal changes known as hyperplasia and dysplasia. Hyperplasia is characterized by an increase in the number of cells in an organ or tissue that appear normal under a microscope. In dysplasia, the cells appear anomalous under a microscope, which may or may not be cancer, hyperplasia, and dysplasia.

1.2 TYPES OF CANCER

Carcinomas: A carcinoma that covers the surface of the internal organs and glands starts in the skin or tissue. Carcinomas form solid tumours; they constitute the most common form of cancer. Examples of carcinomas include prostate, breast, lung, and colorectal cancer.

Sarcomas: A sarcoma begins in the tissues which support the body and connect it. A sarcoma can develop in fat, muscles, nerves, tendons, joints, lymphatic vessels, cartilage, or bone.

Leukaemia: Leukaemia is a blood cancer that starts with changes in healthy blood cells and spreads uncontrollably. Acute lymphocytic leukaemia, chronic lymphocytic leukaemia, acute myeloid leukaemia, and chronic myeloid leukaemia are the four major types of leukaemia.

Lymphomas: Lymphoma is a cancer that starts in the lymph system. The lymph system is a network of vessels and glands that help combat infection. There are two main lymphoma types: Hodgkin lymphoma and non-Hodgkin lymphoma.

Other common types of cancer: The four most common types of cancer are prostate, lungs, colorectal, breast cancer, and brain tumours in men, women, and children, as shown in Figure 1.1.

1.3 CAUSES OF CANCER

The following is a list of major causes, and it is not all-inclusive, as different causes are regularly added with advances in research:

- *Tobacco:* It is the leading cause of cancer-related deaths among men worldwide and increasingly among women. Forms of exposure include active smoking, second-hand smoke breathing (passive or involuntary smoking), and smokeless tobacco. Tobacco causes several types of cancer, including lung, oesophageal, laryngeal, oral, bladder, kidney, stomach, cervical, and colorectal cancer. Total tobacco use mortality in 2005 was estimated at

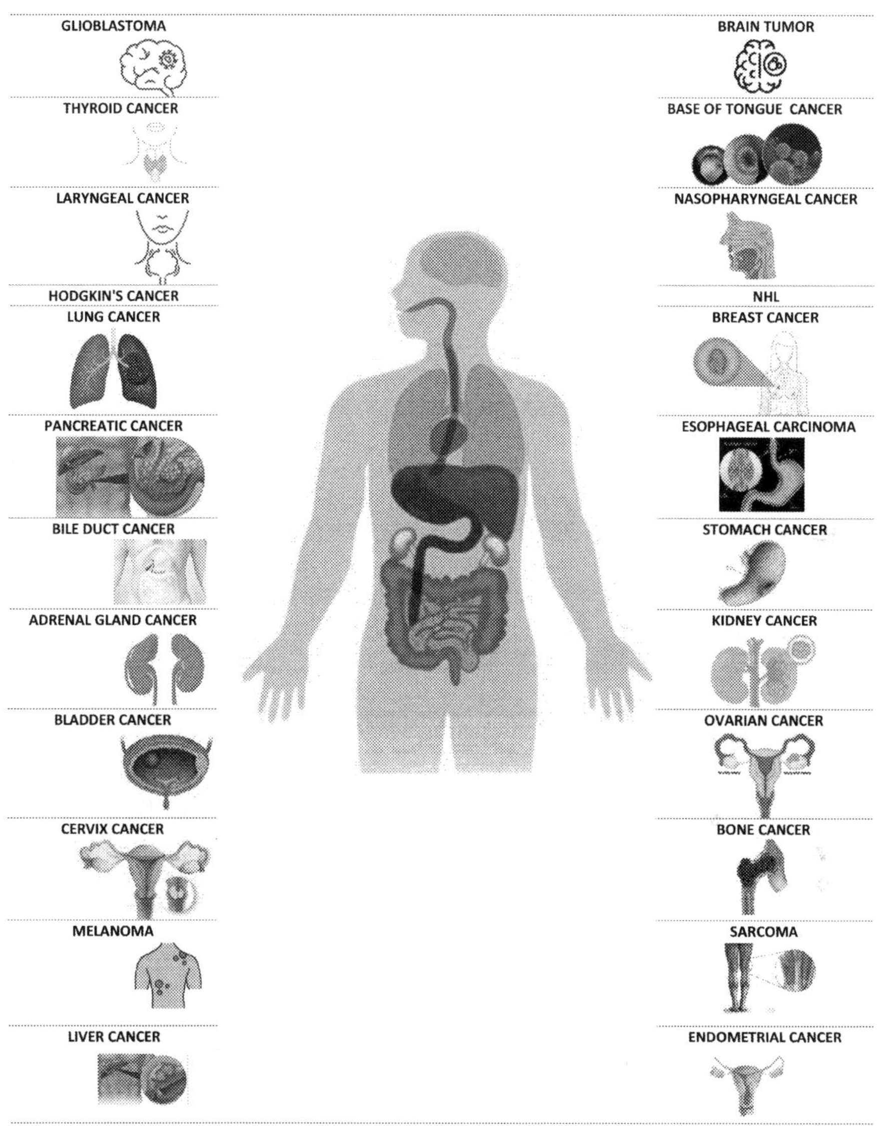

FIGURE 1.1 Common types of cancer.

5.4 million men, including nearly 1.5 million deaths from cancer. If the current rate of tobacco use continues, the overall number of tobacco-related deaths is predicted to increase to nearly 6.4 million in 2015, including 2.1 million cancer deaths. By 2030, it is projected that there will be ~26 million new cancer cases and 17 million cancer deaths per year.

- *Physical Inactivity, Dietary Factors, and Obesity and Overweight:* These play a significant role in causing cancer. Gender norms affect these factors. Since all of these factors are closely intertwined at the individual and contextual levels, it is difficult to quantify the precise impact of each of these risk factors and may underestimate the possible cumulative risk. Overweight and obesity are causally associated with several common types of cancer, including oesophageal, colorectal, postmenopausal women's breast, endometrial, and kidney cancers (WHO, 2003a). Physical inactivity is a significant contributor to the rise in overweight and obesity levels in several areas of the world, and it raises the risk of certain cancers separately. Taken together, increased body mass index and physical inactivity account for an associated 19% of mortality from breast cancer and 26% of mortality from colorectal cancer. Overweight and obesity account for 40% of endometrial (uterine) cancer. Together, overweight, obesity, and physical inactivity account for an estimated 1,59,000 deaths annually from colon and rectum cancer and 88,000 deaths from the breast.

- *Alcohol:* It is a risk factor for many forms of cancer, including oral cavity cancer, pharynx, larynx, oesophagus, liver, colorectal, and breast cancer. The risk of for several types of cancer (e.g., oral cavity, pharynx, larynx, and oesophagus) increases substantially if the person besides drinking heavily is also a heavy smoker. For some forms of alcohol-related cancer, attributable fractions vary between men and women, primarily due to variations in mean intake rates. For example, alcohol is attributable to 22% of men's mouth and oropharynx cancers, whereas the attributable burden drops to 9% in women. There is a similar gender difference for oesophageal and hepatic cancers.
- *Chronic Hepatitis B Virus (HBV) Infection:* Chronic hepatitis causes around 52% of hepatocellular carcinomas worldwide, resulting in nearly 340,000 deaths per year. Around 20% (124,000 deaths) of hepatocellular cancers are caused by infection with hepatitis C virus (HCV). HBV infections interact with aflatoxin exposure (through the consumption of contaminated foods) to increase the risk of hepatic cancer. Both HBV infections and aflatoxin exposure are particularly common in sub-Saharan Africa and some parts of South-East Asia and are thought to be the cause of up to 80% of cases of liver cancer that occur in those regions.
- *Human Papillomavirus (HPV):* HPV is the most common sexually transmitted reproductive tract viral infection in the world, infecting an estimated 660 million people per annum. Nearly all cases of cervical cancer, 90% of anal cancers, and 40% of external genitalia cancers are also estimated to be caused by HPV. HPV also causes cancer of the oral cavity, and oropharynx cancer, and more than ten other forms are causally connected to cervical cancer. The most popular genotypes of high risk account for about 70% of cases of cervical cancer worldwide. However, some regional variations are mainly due to differences in the prevalence of type 18 HPV.
- *Environmental Pollution:* Occupational contamination or carcinogenic contaminants in food, water, and soil accounts for 1–4% of all cancers. Environmental exposure to carcinogenic chemicals may occur by drinking water or indoor and outdoor air contamination. In Bangladesh, 5–10% of all cancer deaths in an arsenic-contaminated area are due to arsenic

exposure to carcinogens, which often happens by chemical degradation of food, such as aflatoxins or dioxins. Indoor air pollution from coal fires doubles the risk of lung cancer, particularly among women who are not smokers. Indoor air pollution from domestic coal fires is responsible for about 1.5% of all deaths from lung cancer worldwide. Coal use is particularly widespread in households in Asia. Over 40 agents, mixtures, and conditions of exposure in the workplace are carcinogenic to humans and are classified as follows:

- *Occupational Carcinogens:* It is well known that occupational carcinogens are causally linked to breast, bladder, larynx, and skin cancer, leukaemia, and nasopharyngeal cancer. Mesothelioma (cancer of the outer lining of the lung or chest cavity) is caused in large measure by work-related asbestos exposure. Occupational cancers are concentrated among specific working population groups, for which the risk of developing a particular form of cancer may be significantly higher than that of the general population. About 20–30% of the male and 5–20% of the working-age female population (people aged 15–64) could have been subjected to lung carcinogens during their working lives, accounting for about 10% of lung cancers worldwide. About 2% of cases of leukaemia are attributable to occupational exposures worldwide.
- *Radiation:* Radiation is the energy that is released by waves or rays. When passing through cells and tissues, ionizing radiation removes electrons from the material (called ionization) leading to cell or tissue injury. Examples of ionizing radiation include medical X-rays and radiation emitted from natural sources such as radon gas and radioactive materials. Ionizing radiation can cause nearly any type of cancer, but especially leukaemia, lung, thyroid, and breast cancer. Natural radiation exposure is largely a result of home radon gas which increases the risk of lung cancer. Non-ionizing radiation includes electromagnetic fields such as those emitted by mobile phones or power lines and ultraviolet (mainly from the sun) radiation, the latter causing chromosome damage. Ultraviolet radiation, like malignant melanomas, is a known cause of skin cancer.
- *Reproductive Factors:* Reproductive factors such as the age of the mother when she first gives birth and the number of births that impact the risk of cancer are not included in this section. Childbirth decisions are typically taken within a broad background of social, family, and adult experiences and are not motivated solely by a desire to minimize cancer risk.
- *Chemical or Toxic Compound Exposures:* Benzene, arsenic, copper, cadmium, vinyl chloride, benzidine, *N*-nitrosamines, nicotine, and cigarette smoke (contains at least 66 known possible carcinogenic chemicals and toxins), arsenic, and aflatoxin.
- *Genetics:* Several specific cancers have been linked to human genes and are as follows: breast, ovarian, colorectal, prostate, skin, and

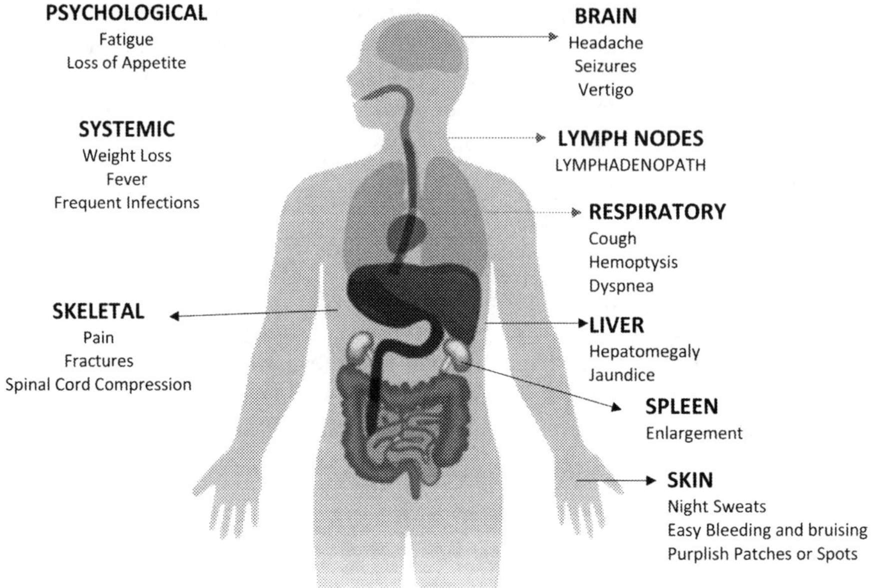

FIGURE 1.2 Common sites and symptoms of cancer.

melanoma; the specific genes and other details go beyond the scope of this chapter, so the reader is referred to the National Cancer Institute for more details on genetics and cancer.

1.4 SYMPTOMS OF CANCER

The American Cancer Society describes six signs and/or symptoms of warning that cancer may occur and that should prompt a person to seek medical attention (Figure 1.2):

- Change in bowel or bladder habits
- A sore throat that is not healing
- Unusual bleeding or discharge (such as nipple secretions or a "sore" that won't cure the oozing material)
- Breast thickening or lumping, testicles, or anywhere else
- Indigestion or difficulty swallowing (usually chronic)
- Obvious modification in size, colour, shape, or thickness

1.5 COMMON RISK FACTORS

- Tobacco use
- Alcohol use
- Dietary considerations
- Poor nutrition, including low fruit and vegetable intake
- Physical inactivity, overweight, radiation exposure, and obesity

1.5.1 Risk Factors of Cancer

Other important cancer risk factors include exposure to the following:

- Physical carcinogens, such as UV and ionizing radiation
- Organic carcinogenic elements, such as benzopyrene, formaldehyde, and aflatoxins (food contaminants), and asbestos fibres
- Biological carcinogens, such as viral, bacterial, and parasitic infections

The interventions aimed at lowering rates in the population of the above-mentioned risk factors would not only decrease the prevalence of cancer but also other diseases that bear these risks. Among the most important cancer risk factors that could be modified are the following:

- Tobacco use is responsible for up to 1.5 million deaths from cancer per year (60% of these deaths occur in low- and middle-income countries); Cancer prevention encompasses a broad spectrum of strategies designed to lower the chance of developing cancer and reduce the morbidity of established cancer. Unfortunately, in many countries, many preventive measures, both cost-effective and inexpensive, have yet to be widely implemented. Together, overweight, obesity, and physical inactivity are responsible for 274,000 deaths from cancer each year.
- Harmful use of alcohol, responsible for 351,000 deaths annually from cancer.
- Infection with sexually transmitted human papillomavirus (HPV), responsible for 2,35,000 deaths from cancer annually.
- Air pollution (outdoor and indoor) is blamed for 71,000 deaths from cancer worldwide.
- Industrial carcinogens are responsible for at least 152,000 deaths from cancer per annum.

1.6 MOLECULAR BIOLOGY OF CANCER

Cancer is a multistep process that requires the accumulation of multiple genetic mutations in a single cell that bestows features characteristic of a neoplastic cell. Typically, tumour cells differ from normal cells in that they exhibit uncontrolled growth. Because features that distinguish tumours from normal cells may be key to understanding neoplastic cell behaviour and may ultimately lead to therapies that can target tumour cells, considerable effort has been directed at identifying the phenotypic characteristics of *in vitro* transformed cells and of tumour cells derived from natural sources.

Molecular biology includes immortality, decreased dependence on growth factors to support proliferation, loss of anchorage-dependent growth, loss of cell cycle control, reduced sensitivity to apoptotic cell death, and increased genetic instability.

1.6.1 Immortality

Normal diploid fibroblasts have a limited capacity to grow and divide both *in vivo* and *in vitro*. Even if provided with optimal growth conditions, *in vitro* normal cells will cease dividing after 50–60 population doublings and then senesce and die. In contrast, malignant cells that have become established in culture proliferate indefinitely and are said to be immortalized. The barrier that restricts the lifespan of normal cells is known as the Hayflick limit and was first described in experiments that attempted immortalization of rodent cells. Normal embryo-derived rodent cells, when cultured *in vitro*, initially divide rapidly. Eventually, however, these cultures undergo a crisis phase during which many of the cells senesce and die. After extended maintenance, however, proliferation in the cultures increases, and cells that can divide indefinitely emerge. The molecular changes that take place during the crisis have revealed at least two important restrictions that must be overcome for cells to become immortalized, and both of these changes occur in natural tumour cells.

One barrier to cellular immortalization is the inability of the DNA replication machinery to efficiently replicate the linear ends of DNA at the 5′ ends, which leads to the shortening of the chromosome. In bacteria, the end replication problem is solved with a circular chromosome. In human cells, the ends of chromosomes are capped with 5–15 kb of repetitive DNA sequences known as telomeres. Telomeres serve as a safety cap of non-coding DNA that is lost during normal cell division without any consequence to the normal function of the cell. However, because telomere length is shortened with each round of cell division, indefinite proliferation is impossible because eventually the inability to replicate chromosomal ends nibbles into DNA containing vital genes. Telomeres seem to be lengthened during gametogenesis because of the activity of an enzyme called telomerase. Telomerase activity has been detected in normal ovarian epithelial tissue. More importantly, telomerase activity is elevated in the tumour tissue but not in the normal tissue from the same patient. This implies that one mechanism by which tumour cells overcome the shortening telomere problem and acquire the capacity to proliferate indefinitely is through abnormal upregulation of telomerase activity. The finding that telomerase activity is found almost exclusively in tumour cells is significant because it suggests that this enzyme may be a useful therapeutic target. Therapies aimed at suppressing telomerase would eliminate a feature essential for tumour cell survival and would be selective.

The second feature of immortalization is loss of growth control by elimination of tumour suppressor activity. Recent evidence suggests that inactivating mutations in both the Rb and p53 tumour suppressor genes occur during the crisis. Both of these genes are discussed in more detail later in this chapter and both function to inhibit cell proliferation by regulating cell cycle progression. Consequently, loss of tumour suppressor function also appears to be a critical event in immortalization.

1.6.2 Decreased Dependence on Growth Factors to Support Proliferation

Cells grown in culture require media supplemented with various growth factors to continue proliferating. In normal human tissues, growth factors are generally produced extracellularly at distant sites and then are either carried through the bloodstream or diffuse to their nearby target cells. The former mode of growth factor stimulation is termed endocrine stimulation, and the latter mode is termed paracrine stimulation. However, tumour cells often produce growth factors that bind to and stimulate the activity of receptors that are also present on the same tumour cells that are producing the growth factor. This results in a continuous self-generated proliferative signal known as autocrine stimulation that drives the proliferation of the tumour cell continuously even in the absence of any exogenous proliferative signal. Autocrine stimulation is manifested as a reduced requirement for serum because the serum is the source of many of the growth factors in the media used to propagate cells *in vitro*.

Because of the prominent role that growth factors and their cognate receptors play in tumour cell proliferation, they have also become favourite therapeutic targets. For example, the epidermal growth factor receptor (EGFR) is known to play a major role in the progression of most human epithelial tumours, and its overexpression is associated with poor prognosis. Consequently, different approaches have been developed to block EGFR activation function in cancer cells, including anti-EGFR blocking monoclonal antibodies (MAb), epidermal growth factor (EGF) fused to toxins, and small molecules that inhibit the receptor's tyrosine kinase activity (RTK). Of these, an orally active anilino quinazoline, ZD1839 ("Iressa"), shows the most promise as an antitumour agent by potentiating the antitumour activity of conventional chemotherapy.

1.6.3 Loss of Anchorage-dependent Growth and Altered Cell Adhesion

Most normal mammalian cells do not grow, but instead they undergo cell death if they become detached from a solid substrate. Tumour cells, however, can frequently grow in suspension or in a semisolid agar gel. The significance of the loss of this anchorage-dependent growth of cancer cells relates to the ability of the parent tumour cells to leave the primary tumour site and become established elsewhere in the body. The ability of cancer cells to invade and metastasize foreign tissues represents the final and most difficult to treat stage of tumour development, and it is this change that accompanies the conversion of a benign tumour to life-threatening cancer.

Metastasis is a complex process that requires the acquisition of several new characteristics for tumour cells to successfully colonize distant sites in the body. Epithelial cells normally grow attached to a basement membrane that forms a boundary between the epithelial cell layers and the underlying supporting stroma separating the two tissues. This basement membrane consists of a complex array of extracellular matrix (ECM) proteins, including type IV collagen, proteoglycans, laminin, and fibronectin, which normally acts as a barrier to epithelial cells. A common feature of tumour cells with metastatic potential is the capacity to penetrate the basement membrane by proteolysis, to survive in the absence of attachment to this substrate, and to colonize and grow in a tissue that may be a foreign relative to the original tissue of origin.

Consequently, metastasis is a multistep process that begins with the detachment of tumour cells from the primary tumour and then penetrates through the basement membrane by degradation of the ECM proteins. This capacity to proteolytically degrade basement membrane proteins is driven, in part, by the expression of matrix metalloproteinases (MMPs). MMPs are a family of enzymes that are either secreted

(MMPs 1–13, 18 –20) or anchored in the cell membrane (MMPs 14–17) (Table 1.1). Regulation of MMPs occurs at several levels: transcription, proteolytic activation of the zymogen, and inhibition of the active enzyme. MMPs are typically absent in normal adult cells, but a variety of stimuli, such as cytokines, growth factors, and alterations in cell–cell and cell–ECM interactions, can induce their expression. The expression of MMPs in tumours is frequently localized to stromal cells surrounding malignant tumour cells. Most of the MMPs are secreted in their inactive (zymogen) form and require proteolytic cleavage to be activated. In some cases, MMPs have been shown to undergo mutual and/or auto-activation *in vitro*.

TABLE 1.1
Matrix Metalloproteinases (MMPs)

MMP	Common Name	Substrates	Cell Surface
1	Collagenase-1, interstitial collagenase	Collagen I, II, III, VII, X, IGFBP	Yes
2	Gelatinase A	Gelatin, collagen I, IV, V, X, laminin, IGFBP, latent TGF-β	Yes
3	Stromelysin-1	Collagen III, IV, V, IX, X, gelatin, E-cadherin, IGFBP, fibronectin, elastin, laminin proteoglycans, perlecan, Hb-EGF, proMMP-13	Unknown
7	Matrilysin	Laminin, fibronectin, gelatin, collagen IV, proteoglycans Fas-L, proMMP-1, Hb-EGF	Yes
8	Collagenase-2, neutrophil collagenase	Collagen I, II, III, VII, X	Unknown
9	Gelatinase B	Collagen I, IV, V, X, gelatin, IGFBP, latent TGF-β	Yes
10	Stromelysin-2	Collagen III, IV, IX, X, gelatin, laminin, proteoglycans, proMMP-1, proMMP-13	Unknown
11	Stromelysin-3	IGFBP, A-1-antiprotease	Unknown
12	Metalloelastase	Elastin, proMMP-13	Unknown
13	Collagenase-3	Collagen I, II, III, II, VII, X, XIV, fibronectin, proMMP-9, tenascin, aggrecan	Unknown
14	Mt1-MMP	Gelatin, collagen I, fibrin, proteoglycans, laminin, fibronectin, proMMP-2	Yes
15	Mt2-MMP	Laminin, fibronectin, proMMP-2, proMMP-13, tenascin	Yes
16	Mt3-MMP	Gelatin, collagen III, fibronectin, proMMP-2	Yes
17	Mt4-MMP	Unknown	Yes
18/19	Rasi-1	Unknown	Unknown
20	Enamelysin	Amelogenin	Unknown

Several lines of evidence implicate MMPs in tumour progression and metastasis. Firstly, MMPs are overexpressed in tumours from a variety of tissues, and the expression of one, matrilysin, is elevated in invasive prostate cancer epithelium. Secondly, reduction of tissue inhibitor of matrix metalloproteinases-1 (TIMP-1) expression in mouse fibroblasts (Swiss 3T3), using antisense RNA technology, increased the incidence of metastatic tumours in immunocompromised mice. Similarly, overexpression of the various MMPs has provided direct evidence for their role in metastasis. Importantly, synthetic MMP inhibitors have also been produced and they lead to a reduction in metastasis in several experimental models of melanoma, colorectal carcinoma, and mammary carcinoma, suggesting a mechanism by which the invasive potential of tumours may be reduced.

Once tumour cells escape through the basement membrane, they can metastasize through two major routes: the blood and lymphatic vessels. Tumours originating in different parts of the body have characteristic patterns of invasion. Some tumours, such as those of the head and neck, spread initially to regional lymph nodes. Others, such as breast tumours, can spread to distant sites relatively early. The site of the primary tumour generally dictates whether the invasion will occur through the lymphatic or blood vessel system. The cells that escape into the vasculature must evade host immune defence mechanisms to be successfully transported to regional or distal locations. Tumour cells then exit blood vessels and escape into the host tissue by again compromising a basement membrane: this time the basement membrane of the blood vessel endothelium. Projections called invadopodium, which contains various proteases and adhesive molecules, adhere to the basement membrane, and this involves membrane components such as laminin, fibronectin, type IV collagen, and proteoglycans. The tumour cells then produce various proteolytic enzymes, including MMPs, which degrade the basement membrane and allow invasion of the host tissue. This process is referred to as extravasation.

The interaction between cells and ECM proteins occurs through cell surface receptors, the best characteristic of which is the fibronectin receptor that binds fibronectin. Other receptors bind collagen and laminin. Collectively, these receptors are called integrins, and their interaction with matrix components conveys regulatory signals to the cell. They are heterodimeric molecules consisting of one of several alpha and beta subunits that may combine in any number of permutations to generate a receptor with distinct substrate preferences. Changes in the expression of integrin subunits are associated with invasive and metastatic cells facilitating invasion by shifting the cadre of integrins to the integrins that preferentially bind the degraded subunits of ECM proteins produced by MMPs. Hence, integrin expression has served as a marker for the invasive phenotype and may be a logical target for novel therapies that interfere with the progress of advanced tumours.

In addition to their role in invasion, the evidence also indicates that MMPs may play a role in tumour initiation and in tumourigenicity. Expression of MMP-3 in normal mammary epithelial cells led to the formation of invasive tumours. A proposed mechanism for this initiation involves the ability of MMP-3 to cleave E-cadherin. E-cadherin is a protein involved in a cell–cell adhesion together with other proteins such as β-catenin and actinin. Loss of E-cadherin function is known to lead to tumourigenicity and invasiveness because of the loss of cellular adhesion.

Interestingly, inhibition of MMP-7 and MMP-11, using antisense approaches, did not affect invasiveness or metastatic potential *in vitro*. However, tumourigenicity was altered. Matrilysin, MMP-7 messenger RNA (mRNA), are present in benign tumours and malignant tumour cells of the colon. The relative level of matrilysin expression correlates with the stage of tumour progression.

1.6.4 Cell Cycle and Loss of Cell Cycle Control

Proliferation is a complex process consisting of multiple subroutines that collectively bring about cell division. At the heart of proliferation is the cell cycle, which consists of many processes that must be completed in a timely and sequence-specific manner. Accordingly, the regulation of cell cycle events is a multifaceted affair and consists of a series of checks and balances that monitor nutritional status, cell size, presence or absence of growth factors, and integrity of the genome. These cell cycle regulatory pathways and the signal transduction pathways that communicate with them are populated with oncogenes and tumour suppressor genes.

Cell division is divided into four phases: G1, S, G2, and M. The entire process is punctuated by two spectacular events: the replication of DNA during the S phase and chromosome segregation during the mitosis or M phase. Of the four cell cycle phases, three can be assigned to replicating cells and only during the G1 phase, and a related quiescent phase, G0, is non-replicative in nature. Normal cycling cells that cease to proliferate enter the resting phase, or G1 is strongly dependent on the presence of growth factors and nutrients. Hence, the conditions that lead to exit from G1 and entry into S are tightly regulated and are frequently regulated in neoplastic cells that exhibit uncontrolled proliferation.

Movement through the cell cycle is controlled by two classes of cell cycle proteins—cyclins and cyclin-dependent kinases (CDKs), which physically associate to form a protein kinase that drives the cell cycle forward. At least 8 cyclins and 12 CDKs have been identified in mammalian cells. The name "cyclin" derives from the characteristic rise and fall in abundance of cyclin B as cells progress through the cell cycle. The accumulation of cyclin proteins occurs through the cell cycle–dependent induction of gene transcription, but the elimination of cyclins occurs by carefully regulated degradation that is enabled through protein sequence tags known as destruction boxes and PEST sequences. Although not all the cyclin types exhibit this oscillation in protein quantity, those cyclins that play key roles in progression through the cell cycle (cyclins E, A, and B) are most abundant during discrete phases of the cell cycle. Cyclin D1 is synthesized during the G1 phase just before the restriction point and plays an important role in the regulation of the R point. Cyclin E is most abundant during the late G1 and early S phases and is essential for exit from the G1 phase and progression into the S phase. Elevated levels of these two G1 cyclins can result in uncontrolled proliferation. Indeed, both cyclin D1 and cyclin E are overexpressed in some tumour types, suggesting that the cyclins and other components of the cell cycle may be useful therapeutic targets (Figure 1.3).

The second component of the enzyme complex is CDK that, as the name implies, requires an associated cyclin to become active. At least 12 protein kinases have been isolated from humans, *Xenopus*, and *Drosophila*, and are numbered according to

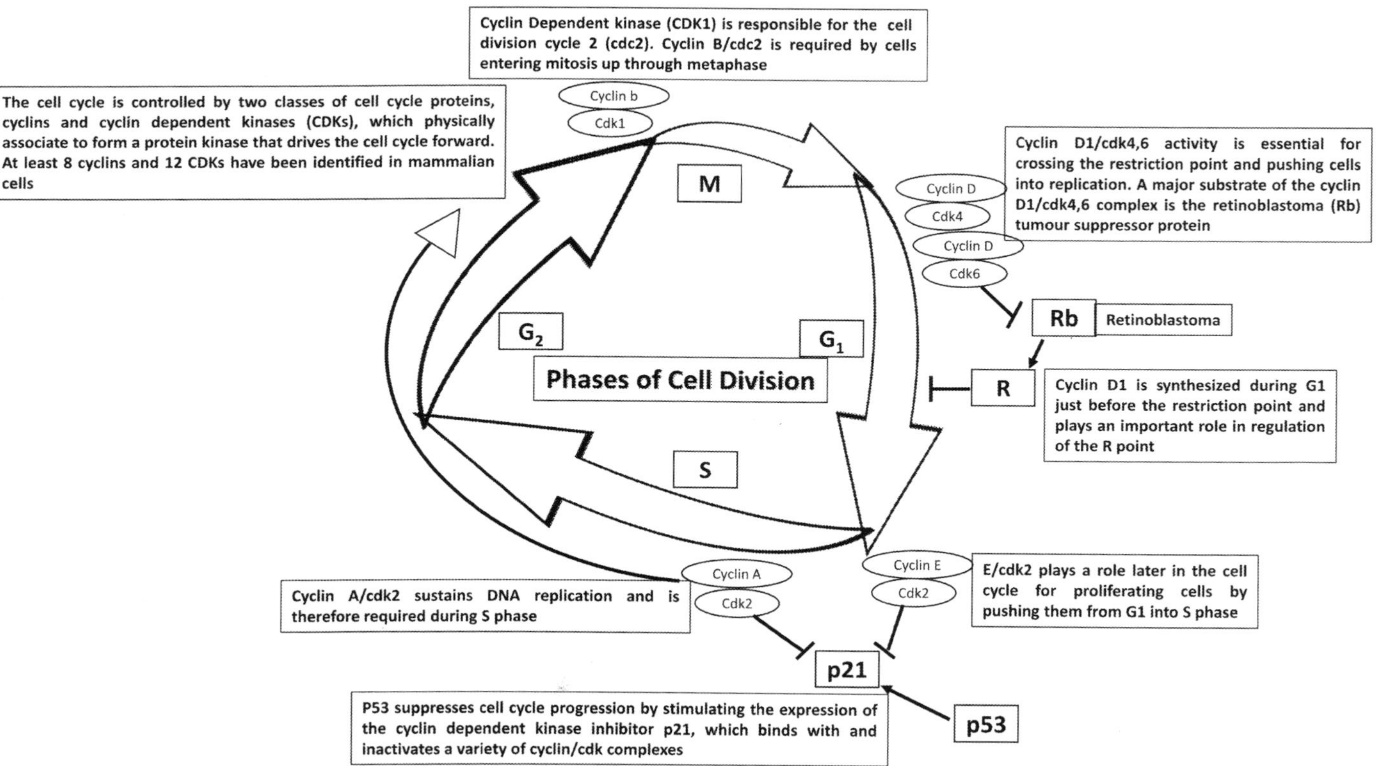

FIGURE 1.3 Schematic representation of cell cycle.

a standardized nomenclature beginning with CDK1, which for historical reasons, is most frequently referred to as cell division cycle 2 (cdc2). Unlike the cyclins, the abundance of the CDK proteins remains relatively constant throughout the cell cycle. Instead, their activity changes during different phases of the cell cycle in accordance with whether or not an activating cyclin is present and whether or not the kinase itself is appropriately phosphorylated. Both cyclins and CDKs are highly conserved from yeast to man and function similarly, suggesting that the cell cycle is controlled by a universal cell cycle engine that operates through the action of evolutionarily conserved proteins. Hence, drug discovery studies aimed at identifying agents that regulate the cell cycle may be performed in model organisms, such as yeast, *Caenorhabditis elegans*, and *Drosophila*, with some assurance that the targeted mechanisms will also be relevant to humans.

It is now clear that specific cyclin/CDK complexes are required during specific stages of the cell cycle. Cyclin D1/cdk4,6 activities are essential for crossing the restriction point and pushing cells into replication. A major substrate of the cyclin D1/cdk4,6 complex is the retinoblastoma (Rb) tumour suppressor protein, which, when phosphorylated by this kinase complex, is inactivated. This frees the cell from the restrictions on cell proliferation imposed by the Rb protein. It is this event that is believed to be decisive in the stimulation of resting cells to undergo proliferation. Cyclin E/cdk2 plays a role later in the cell cycle for proliferating cells by pushing them from the G1 phase into the S phase. Cyclin E is overexpressed in some breast cancers which may enhance the proliferative capacity of tumour cells. Cyclin A/cdk2 sustains the DNA replication and is therefore required during the S phase. Cyclin B/cdc2 is required by cells entering mitosis up through metaphase. At the end of metaphase, cyclin B is degraded and cdc2 becomes inactivated, allowing mitotic cells to progress into anaphase and complete mitosis. Sustaining the activity of cyclin B/cdc2 causes cells to arrest in metaphase. Hence, it is the collective result brought about by the activation and deactivation of cyclin/CDK complexes that pushes proliferating cells through the cell cycle.

Superimposed on the functions of the cell cycle engine is a complex network of both positive and negative regulatory pathways. Important negative regulators are the cyclin-dependent kinase inhibitors (CKIs). There are two families of CKIs: the Cip/Kip family and the INK4 family. The Cip/Kip family consists of three members: p21/Cip1/waf1/Sdi1, p21/Kip1, and p57/Kip2. All of the proteins in this family have broad specificity and can bind to and inactivate most of the cyclin–CDK complexes that are essential for progression through the cell cycle. p21waf1, the first discovered and best-characterized member of the Cip/Kip family, is stimulated by the p53 tumour suppressor protein in response to DNA damage and halts cell cycle progression to allow for DNA repair. The INK4 family of CKIs contains four member proteins: p16/INK4a, p15/INK4b, p18/INK4c, and p19/INK4d. Unlike the Cap/Kip family, the INK4 proteins have restricted binding and associate exclusively with cdk4/6. Consequently, their principal function is to regulate the cyclin D1/cdk4/6 activity, and, therefore, the phosphorylation status of the Rb tumour suppressor. p16/INK4a is itself a tumour suppressor that is frequently mutated in melanoma. Indeed, at least one component of the p16/cyclin D1/Rb pathways is either mutated or deregulated in some fashion in over 90% of lung cancers, emphasizing the importance of this pathway in regulating the tumour cell proliferation.

Transit through the cell cycle is regulated by two types of controls. In the first type, the cumulative exposure to specific signals, such as growth factors, is assessed and if the sum of these signals satisfies the conditions required by the R point, proliferation ensues. In the second type, feedback controls or checkpoints monitor whether the genome is intact and whether previous cell cycle steps have been completed. At least five cell cycle checkpoints have been identified, two that monitor the integrity of the DNA and halt cell cycle progression in either G1 or G2, one that ensures DNA synthesis has been completed before mitosis begins, one that monitors completion of mitosis before allowing another round of DNA synthesis, and one that monitors chromosome alignment on the equatorial plate before initiation of anaphase. Of these, the two checkpoints that monitor the integrity of DNA have been the most extensively studied, and as might be expected, these checkpoints and the genes that enforce them are critically important for the response that cells mount to genotoxic stresses. Abrogation of checkpoints leads to genomic instability and an increased mutation frequency. Progress in elucidating the mechanisms of checkpoint function reveals that a number of checkpoint genes are frequently mutated in human cancers. For example, the p53 tumour suppressor functions as a cell cycle checkpoint that halts cell cycle progression in G1 by inducing the expression of the p21waf1 gene in the presence of damaged DNA. The p53 gene is frequently mutated in human cancers and, consequently, most tumour cells lack the DNA damage–induced p53-dependent G1 checkpoint, increasing the likelihood that mutations will be propagated in these cells. Because p53 also promotes apoptosis, the lack of p53 in these cells also makes them more resistant to DNA damage–induced apoptosis. Because most chemotherapeutic agents kill cells through DNA damage–induced apoptosis, tumour cells with mutant p53 are also more resistant to conventional therapies.

1.6.5 Apoptosis and Reduced Sensitivity to Apoptosis

Apoptosis is a genetically controlled form of cell death that is essential for tissue remodelling during embryogenesis and for the maintenance of the homeostatic balance of cell numbers later in adult life. The importance of apoptosis to human disease comes from the realization that disruption of the apoptotic process is thought to play a role in diverse human diseases ranging from malignancy to neurodegenerative disorders. Because apoptosis is a genetically controlled process, much effort has been spent on identifying these genetic components to better understand the apoptotic process as well as to identify potential therapeutic targets that might be manipulated in disease conditions where disruption of apoptosis occurs.

Although multiple forms of cell death have been described, apoptosis is characterized by morphological changes, including cell shrinkage, membrane blebbing, chromatin condensation, nuclear fragmentation, loss of microvilli, and extensive degradation of chromosomal DNA (Figure 1.4). In general, the apoptotic program can be subdivided into three phases: the initiation phase, the decision/effector phase, and the degradation/execution phase. In the initiation phase, signal transduction pathways that are responsive to external stimuli, such as death receptor ligands, or to

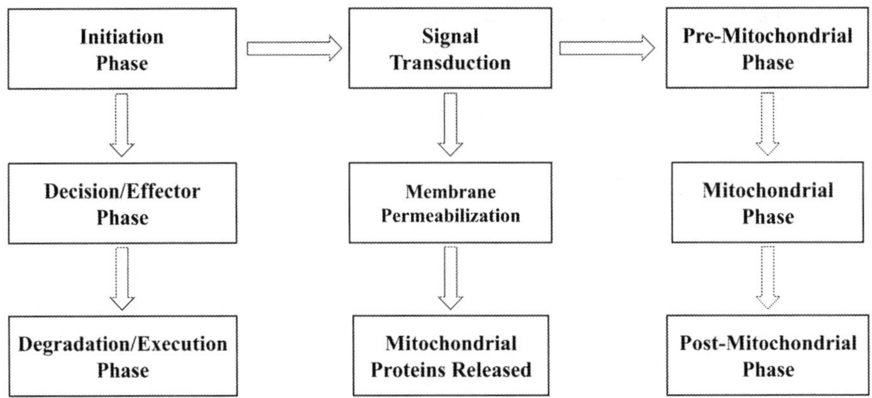

FIGURE 1.4 Schematic representation of mitochondria-mediated apoptosis.

internal conditions, such as that produced by DNA damage, are activated. During the ensuing decision/effector phase, changes in the mitochondrial membrane occur that result in disruption of the mitochondrial membrane potential and ultimately loss of mitochondrial membrane integrity. A key event in the decision/effector phase is the release of cytochrome c into the cytoplasm and the activation of proteases and nucleases that signal the onset of the final degradation/execution phase. An important concept in understanding apoptosis is that the mitochondrion is a key target of apoptotic stimuli and disruption of mitochondrial function is central to subsequent events that lead to the degradation of vital cellular components. Of the signal transduction pathways that initiate apoptosis, the best understood at the molecular level involves the death receptors, including Fas/cluster of differentiation 95 (CD95), tumour necrosis factor receptor 1 (TNFR1), and death receptors 3, 4, and 5 (DR 3, 4, 5). All death receptors share an amino acid sequence known as the death domain (DD) that functions as a binding site for a specific set of death-signalling proteins. Stimulation of these transmembrane receptors can be induced by interaction with its cognate ligand or by binding to an agonistic antibody, which results in receptor trimerization and recruitment of intracellular death molecules and stimulation of downstream signalling events. Here death receptors are classified as either CD95-like (Fas/CD95, DR4, and DR5) or TNFR1-like (TNFR1, DR3, and DR6) based on the downstream signalling events that are induced as a consequence of receptor activation.

Activation of Fas/CD95 leads to clustering and recruitment of the Fas-associated death domain (FADD; sometimes called Mort1) to the Fas/CD95 intracellular DD. FADD contains a C-terminal DD that enables it to interact with trimerized Fas receptor as well as an N-terminal death effector domain (DED), which can associate with the prodomain interacting of the serine protease, caspase-8. This complex is referred to as the death-inducing signalling complex (DISC). As more procaspase-8 is recruited to this complex, caspase-8 undergoes trans-catalytic cleavage to generate active protease. Activation of TNFR1-like death receptors results in similar events except that the first protein to be recruited to the activated receptor is the TNFR-associated death domain (TRADD) adaptor protein that subsequently recruits FADD and procaspase-8. Signalling through the TNFR1-like receptors is more complex and includes the recruitment of other factors that do not interact with Fas/CD95. For example, TRADD also couples with the receptor-interacting protein (RIP), which links stimulation of TNFR1 to signal transduction mechanisms, leading to activation of nuclear factor-kappa B (NF-κB). Because RIP does not interact with Fas/CD95, this class of receptors does not activate NF-κB.

The critical downstream effectors of death receptor activation are the caspases, and these are considered the engine of apoptotic cell death. Caspases are a family of cysteine proteases with at least 14 members. They are synthesized in the cells as inactive enzymes that must be processed by proteolytic cleavage at aspartic acid residues. These cleavage sites are between the N-terminal pro-domain, the large P20, and the small P10 domains. The activated proteases cleave other proteins by recognizing an aspartic acid residue at the cleavage site and are consistent with an auto- or trans-cleavage processing mechanism for activation when recruited to activated death receptors.

Importantly, biochemical studies support the notion of a caspase hierarchy that consists of initiators and effectors that are activated in a cascade fashion. Initiator caspases such as caspase-8 and caspase-9 are activated directly by apoptotic stimuli and function, in part, by activating effector caspases such as caspase-3, caspase-6, and caspase-7 by proteolytic cleavage. It is the effector caspases that result in highly specific cleavage of various cellular proteins and the biochemical and morphological degradation associated with apoptosis.

In contrast to death receptor–mediated apoptosis that functions through a well-defined pathway, mediators of stress-induced apoptosis such as growth factors, cytokines, and DNA damage activate diverse signalling pathways that converge on the mitochondrial membrane. Many proapoptotic agents have been shown to disrupt the mitochondrial membrane potential (ATm), leading to an increase in membrane permeability and the release of cytochrome c into the cytosol. Cytochrome c release is a common occurrence in apoptosis and is thought to be mediated by the opening of the permeability transmembrane pore complex (PTPC), a large multiprotein complex that consists of at least 50 different proteins. The cytosolic cytochrome c interacts with apoptosis-activating factor-1 (Apaf-1), dATP/ATP, and procaspase-9 to form a complex known as the apoptosome. Cytochrome c and dATP/ATP stimulate Apaf-1 self-oligomerization and trans-catalytic activation of procaspase-9 to the active enzyme. Active caspase-9 activates effector caspase-3 and caspase-7 and leads to the cellular protein degradation characteristic of apoptosis.

As the release of cytochrome c can have dire consequences for the viability of the cell, its release is tightly regulated. Indeed, a whole family of proteins, of which B-cell lymphoma-2 (Bcl-2) is the founding member, that share homology in regions called the Bcl-2 homology domains are dedicated to the regulation of cytochrome c release from the mitochondria. Both positive regulators (Bax, Bak, Bik, and Bid), which promote apoptosis, and negative regulators (Bcl-2 and Bcl-XL), which suppress apoptosis, act by regulating the permeability of the mitochondrial membrane to cytochrome c. Bcl-2 family members have been found in both the cytosol and associated with membranes. Bax is normally found in the cytosol, but subcellular localization changes during apoptosis. Bax has been shown to insert into the mitochondrial membrane where, because of its structure that is like other pore-forming proteins, it is thought to promote the release of cytochrome c. Bcl-2 functions by inhibiting the insertion of Bax into the mitochondrial membrane. Hence, a key factor that determines whether a cell will undergo apoptosis is the ratio of proapoptotic to antiapoptotic Bcl-2 family proteins.

Because apoptosis serves to eliminate cells with high neoplastic potential, cancer cells have evolved to evade apoptosis primarily through two mechanisms. In the first of these, Bcl-2, which suppresses apoptosis, is overexpressed. The Bcl-2 oncogene was first identified as a breakpoint in chromosomal translocations that frequently occurred in B-cell-derived human tumours. Characterization of the rearrangements revealed that the Bcl-2 gene is overexpressed by virtue of being placed adjacent to the powerful IgH promoter. Cloning of the Bcl-2 gene and overexpression in cells of B-cell lineage reduced the sensitivity of these cells to apoptosis and allowed them to survive under conditions that ordinarily caused normal cells to die.

The second mechanism that provides cancer cells with resistance to apoptosis is the suppression of the Fas receptor. As with other receptors, mutations can occur in either the ligand binding domain or in the intracellular domain interfering with the activation of the death signalling pathway. More recently, a novel mechanism for suppressing Fas receptor activation has been identified in which cancer cells synthesize decoy receptors to which ligands can bind but are unable to induce apoptosis (Figure 1.5).

1.6.6 Increased Genetic Instability

A hallmark of tumour cells is genetic instability that is manifested at the chromosomal level as either aneuploidy (the gain or loss of one or more specific chromosomes) or polyploidy (the accumulation of an entire extra set of chromosomes). Acquisition of extra chromosomes is one mechanism by which extra copies of a growth-promoting gene can be acquired by cancer cells, providing them with a selective growth advantage. Structural abnormalities are also common in advanced tumours that lead to various types of chromosomal rearrangements. Translocations and random insertion of genetic material into one chromosome from another can place genes that are not normally located adjacent to one another in proximity, usually leading to abnormal gene expression. Some of these rearrangements are routinely observed in some cancers such as in Burkitt's lymphoma where rearrangements involving chromosome 8 and 14 lead to abnormal expression of the c-myc proto-oncogene as a consequence of being placed adjacent to the immunoglobulin heavy chain promoter.

In chronic myelogenous leukaemia (CML), an abnormal chromosome known as the Philadelphia chromosome results from a translocation involving chromosomes 9 and 22. The genes for two unrelated proteins, c-Abl and Bcr, a tyrosine kinase and a GTPase-activating protein (GAP), are spliced together, forming a chimeric protein that results in a powerful and constitutively active kinase that drives the proliferation of the cells in which it is expressed.

Other forms of genetic instability include gene amplification. Under normal conditions, all DNA within the cell is replicated uniformly and only once per cell cycle. However, in cancer cells, some regions of a chromosome can undergo multiple

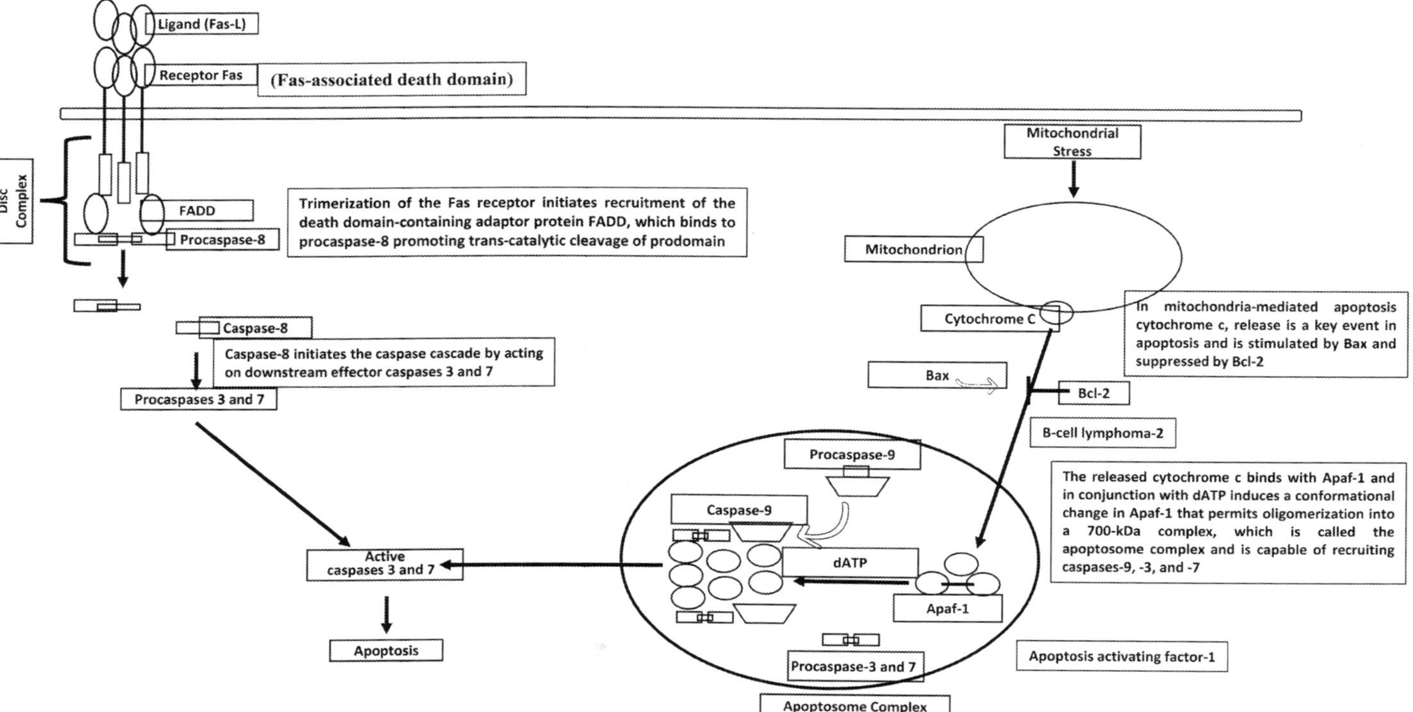

FIGURE 1.5 Schematic representation of biology of cancer through apoptosis.

rounds of replication such that multiple copies of a growth-promoting gene(s) are obtained. These can result in chromosomes with regions of DNA that stain uniformly during karyotype analysis of a tumour cell or in the production of extra chromosomal DNA-containing bodies known as double minute chromosomes. A typical example of this type of amplification targets the N-myc gene, which is amplified in ~30% of advanced neuroblastomas.

More subtle changes at the sequence level affecting growth-controlling genes are also common in human tumours. Mutations can occur because of either defects in DNA repair or decreased fidelity during the DNA replication. The components of these pathways are critical for the maintenance of genome integrity and inherited mutations in the genes of DNA repair proteins and proteins that repair replicated DNA explain some inherited cancer-prone syndromes.

1.6.7 Angiogenesis

Without the production of new blood vessels, tumour growth is limited to a volume of a few cubic millimetres by the distance that oxygen and other nutrients can diffuse through tissues. As tumour size increases, intra-tumoural O_2 levels fall and the centre of the mass becomes hypoxic, leading to upregulation of the hypoxia-inducible factor (HIF1). HIF1 is a heterodimeric transcription factor composed of a constitutively expressed HIF1-β subunit and an O_2-regulatable HIF1-α subunit. Under harmonic conditions, levels of HIF1 are kept low through the actions of the VHL tumour suppressor protein, which functions as a ubiquitin ligase that promotes degradation through a proteasome-mediated pathway. An important transcriptional target of HIF1 is the VEGF growth factor, which in conjunction with other cytokines, induces neovascularization of tumours and allows them to grow beyond the size limitation imposed by oxygen diffusion. This increased production of proangiogenic factors and reduction of anti-angiogenic factors is known as the "angiogenic switch" and is a significant milestone in tumourigenesis that leads to the development of more lethal tumours.

Angiogenesis is the sprouting of capillaries from pre-existing vessels during embryonic development and is almost absent in adult tissues with the exception of transient angiogenesis during the female reproductive cycle and wound-healing, and the soluble factor that plays a critical role in promoting angiogenesis is vascular endothelial growth factor (VEGF). VEGF was first implicated in angiogenesis when it was identified as a factor secreted by tumour cells, which caused normal blood vessels to become hyperpermeable. The following evidence supports the role of VEGF in tumour angiogenesis.

1. VEGF is present in almost every type of human tumour. It is especially high in concentration around tumour blood vessels and in hypoxic regions of the tumour.
2. VEGF receptors are found in blood vessels within or near tumours.
3. Monoclonal neutralizing antibodies for VEGF can suppress the growth of VEGF-expressing solid tumours in mice. These lack any effect in cell culture where angiogenesis is not needed.

Ferrara and Henzel identified VEGF as a growth factor capable of inducing the proliferation of endothelial cells but not fibroblasts or epithelial cells. Inhibition of one of the identified VEGF receptors, FLK1, inhibits the growth of a variety of solid tumours. Similarly, the injection of an antibody to VEGF strongly suppresses the growth of solid tumours of the subcutaneously implanted human fibrosarcoma cell line HT-1080.

There are several forms of VEGF that seem to have different functions in angiogenesis. These isoforms are VEGF, VEGF-B, VEFG-C, and VEGF-D. VEGF-B is found in a variety of normal organs, particularly the heart and skeletal muscle. It can form heterodimers with VEGF and can affect the availability of VEGF for receptor binding. VEGF-D seems to be regulated by c-fos and is strongly expressed in the fatal lung. However, in the adult, it is mainly expressed in skeletal muscle, heart, lung, and intestine. VEGF-D is also able to stimulate endothelial cell proliferation. VEGF-C is about 30% homologous to VEGF. Unlike both VEGF and VEGF-B, VEGF-C does not bind to heparin. It can increase vascular permeability and stimulate the migration and proliferation of endothelial cells, although at a significantly higher concentration than VEGF. VEGF-C is expressed during embryonic development where lymphatics sprout from venous vessels. It is also present in adult tissues and may play a role in lymphatic endothelial differentiation. Flt-4, the receptor for VEGF-C, is expressed in angioblasts, veins, and lymphatics during embryogenesis, but it is mostly restricted to the lymphatic endothelium in adult tissues. Because of these expression patterns, VEGF-C and Flt-4 may be involved in lymph angiogenesis. This is the process of lymphatic generation. Lymphatic vasculature is very important because of its involvement in lymphatic drainage, immune function, inflammation, and tumour metastasis. Other cytokines and growth factors also play an important role in promoting angiogenesis. Some of these act directly on endothelial cells, whereas others stimulate adjacent inflammatory cells. Some can cause migration but not division of endothelial cells such as anisotropic, macrophage-derived factor and TNFα, or stimulate proliferation such as EGF and acidic and basic fibroblast growth factors (aFGF and bFGF), transforming growth factor β (TGF-β) and VEGF. Tumours secrete these factors, which stimulate endothelial migration, proliferation, proteolytic activity, and capillary morphogenesis.

Several angiogenic factors have been identified that can be secreted from tumours. Many of these are growth factors that are described as heparin-binding growth factors. Specifically, these include VEGF, FGFs, TGF-β, and the hepatocyte growth factor (HGF). The binding of these factors to heparin sulphate proteoglycans (HSPG) may be a mechanism for bringing the growth factors to the cell surface and presenting them to their appropriate receptors in the proper conformation. This facilitates the interaction between the growth factors and receptors. Studies have shown that tumour growth is adversely affected by agents that block angiogenesis but is stimulated by factors that enhance angiogenesis.

Angiogenesis may be useful as a prognostic indicator. Tumour sections can be stained immune chemically for angiogenic determinants, such as VEGF, to determine the density of vasculature within the tumour, and there is a strong correlation between high vessel density and poor prognosis. This correlation implies a relationship between angiogenesis and metastasis.

1.7 CANCER DISEASE AS A METABOLIC DISORDER: MANAGEMENT AND PREVENTION

Cancer prevention is an essential part of fighting cancer. Unfortunately, other prevention steps are yet to be broadly adopted in many countries, both cost-effective and affordable.

1.7.1 Diagnosis

The first important stage in cancer treatment is the confirmation of a diagnosis based on clinical analysis. It is possible to extract a tumour sample by conducting a biopsy or aspiration that may require an operation such as an image-guided procedure or endoscopy. Pathology and laboratory medicine facilities are important for the correct examination and evaluation of medical samples, thereby directing the patient's diagnosis, care, and management. Determining the point, that is, the degree of tumour spread from the primary site, is crucial. Staging is used to guide treatment options, and to estimate the prognosis of an individual.

Diagnosing cancer at its earliest stages often provides the best chance of getting a cure for a few cancers, studies show screening tests can save lives by early diagnosis of cancer. Screening tests are recommended for other cancers only for those with an increased risk. A variety of medical organizations and patient advocacy groups have cancer screening recommendations and guidelines.

1.7.2 Approaches to Diagnosing Cancer

- *Physical Examination:* The doctor may feel lumps in areas of your body that might indicate a tumour. He or she may look for abnormalities during a physical examination, such as changes in skin colour or an organ enlargement that may indicate the presence of cancer.
- *Laboratory Tests:* Laboratory tests, such as urine and blood tests, may help your doctor identify cancer-causing abnormalities. For example, a common blood test called complete blood count can reveal an unusual number or type of white blood cells in people with leukaemia.
- *Imaging Tests:* Imaging tests allow non-invasive examination of your bones and internal organs by your doctor. Imaging instruments used in cancer diagnosis include, among others, a computerized tomography (CT) scan, bone scan, magnetic resonance imaging (MRI), positron emission tomography (PET) scan, ultrasound, and X-ray scan.
- *Biopsy:* During a biopsy, your doctor will collect a sample of cells for laboratory testing. There are various ways to gather a sample. Which biopsy procedure is right for you depends on your cancer type and where it is located? In most cases, a biopsy is the only way to diagnose cancer once and for all.
- In the laboratory, doctors look at cell samples under the microscope. Normal cells look uniform, with similar sizes and orderly organization. Cancer cells look less orderly, with varying sizes and without apparent organization.

1.8 TREATMENT OF CANCER

Treatment for cancer requires a thorough evaluation of evidence-based treatments, which can include chemotherapy, radiotherapy, and clinical treatment of more than one of the most effective medical modalities. Considering the resources available, the determination will be focused on evidence of the best treatment required. It is best to share decisions and take into account patient considerations like client desires.

Both types of cancer therapy may have a major psychosocial and financial impact on a patient and their family, which can be taken into account when developing approaches to improve access to and affordability for cancer treatment.

The WHO list of important medicines was revised in 2015 to include 30 cytotoxic and adjuvant drugs (anti-cancer drugs) that are part of proven scientifically effective treatment regimens. This list was examined for efficacy, safety, and quality, and comparative cost-effectiveness assessments were conducted to generate these important trees with other alternatives of the same class of medications. Cancer care programmes require a robust community network to ensure high-quality, reliable, secure, and affordable treatment for all cancer patients.

1.8.1 Goals of Cancer Treatment

- *Cure:* The aim of the therapy is to heal your cancer, allowing you to live a normal lifespan.
- *Primary Treatment:* Cancer therapy may be used as a primary treatment; surgery is the most effective primary cancer treatment for the more common cancers. When your cancer is especially prone to radiation therapy or chemotherapy, the main cure may be one of those treatments.
- *Adjuvant Treatment:* The goal of adjuvant therapy is to destroy any cancer cells that might linger following primary treatment to decrease the risk of recurrence of cancer. This may be used as an adjuvant drug for any cancer diagnosis. Common adjuvant therapies include chemotherapy and hormone therapy.
- *Palliative Treatment:* Palliative therapy may help alleviate the side effects of surgery, or cancer-caused signs and symptoms. Chemotherapy, surgery, radiation, and hormone therapy can all be used to relieve signs and symptoms. Symptoms such as nausea and shortness of breath may be relieved with medicine. Palliative treatment may be used in conjunction with other treatments designed to cure your cancer.

1.8.2 Tools to Treat Cancer

- *Surgery:* The aim of surgery is to remove cancer or as much cancer as possible.
- *Chemotherapy:* Chemotherapy uses pharmaceutical medicines to destroy cancer cells.
- *Radiation Therapy:* Radiation therapy uses powerful beams of energy, such as X-rays, to kill cancer cells. Radiation treatment may come from a machine outside your body (external radiation from the beam), or may be placed inside your body (brachytherapy).

- *Bone Marrow Transplant:* Bone marrow transplant is also known as stem cell transplant. The bone marrow is the substance that makes the blood cells within the body. A bone marrow transplant can use your own cells or cells from a donor.
- *Immunotherapy:* Immunotherapy, also called cellular medicine, uses the immune response in the body to fight cancer. Cancer can live unnoticed in your body since it is not regarded as an attacker by the immune system. Immunotherapy can help your immune system "see" cancer and attack it.
- *Hormone Therapy:* Some types of cancer are fuelled by the hormones in your body. Breast cancer and prostate cancer are examples. Removing those hormones or blocking their effects from the body can cause cancer cells to stop growing.
- *Targeted Drug Therapy:* Targeted drug therapy focuses on particular abnormalities that enable them to live inside cancer cells.
- *Clinical Trials:* Clinical experiments are research where alternative methods of treating cancer are tested. Thousands of clinical trials are pending for cancer.

1.9 CONCLUSION

Cancer is a disorder in which cells develop abnormally (usually derived from a single abnormal cell). The cells have lost their usual control mechanisms, allowing them to multiply indefinitely, invade surrounding tissues, migrate to distant areas of the body, and facilitate the formation of new blood vessels from which they obtain nutrients. The most recent developments in cancer research have given way to a whole new viewpoint on how to treat cancer. These breakthroughs have resulted from a greater understanding of cancer's molecular basis. Some of the older therapies are still useful, although they have some disadvantages. Surgery and radiation, for example, are successful, but they only treat a single localized area of cancer. Chemotherapy can treat cancer cells that have spread across the body, but it comes with a slew of dangerous side effects. Many of these therapies are still in use today and will likely continue to be used in the future, but they will not be the only treatments available. The development of new molecular based or immunological therapies is still in its early stages.

LIST OF ABBREVIATIONS

aFGF, bFGF Acidic and basic fibroblast growth factors
Apaf-1 Apoptosis activating factor-1
Bcl-2 B-cell lymphoma-2
cdc2 Cell division cycle 2
CDKs Cyclin-dependent kinases
CML Chronic myelogenous leukaemia
CT Computerized tomography
DED Death effector domain
DISC Death-inducing signalling complex
DNA Deoxyribose nucleic acid
ECM Extracellular matrix

EGFRE	pidermal growth factor receptor
FADD	Fas-associated death domain
GAP	GTPase activating protein
HBV	Hepatitis B virus
HCV	Hepatocellular virus
HGF	Hepatocyte growth factor
HIF1	Hypoxia-inducible factor
HPV	Human papillomavirus
HSPG	Heparin sulphate proteoglycans
Mab	Monoclonal antibodies
MMPs	Matrix metalloproteinases
PET	Positron emission tomography
PTPC	Permeability transmembrane pore complex
RNA	Ribonucleic acid
RTK	Receptor's tyrosine kinase
TGF-β	Transforming growth factor-β
TIMP-1	Tissue inhibitor of matrix metalloproteinases-1
TNFR1	Tumour necrosis factor receptor 1
TRADD	TNFR-associated death domain
VEGF	Vascular endothelial growth factor
WHO	World Health Organization

BIBLIOGRAPHY

Achen, M. G., Jeltsch, M., Kukk, E., Makinen, T., Vitali, A., Wilks, F., Alitalo, K., & Stacker, S. A. (1998). Vascular endothelial growth factor D (VEGF-D) is a ligand for the tyrosine kinases VEGF receptor 2 (Flk1) and VEGF receptor 3 (Flt4). *Proceedings of the National Academy of Sciences of the United States of America*, *95*, 548–553.

Asano, M., Yukita, A., Matsumoto, T., Kondo, S., & Suzuki, H. (1995). Inhibition of tumor growth and metastasis by an immunoneutralizing monoclonal antibody to human vascular endothelial growth factor/vascular permeability factor121.*Cancer Research*, *55*, 5296–5301.

Chang, R. (2002). Bioactive polysaccharides from traditional Chinese medicine herbs as anticancer adjuvants. *Journal of Alternative and Complementary Medicine*, *8*(5), 559–565.

Chen, H. S., Tsai, Y. F., Lin, S., Lin, C.-C., Khoo, K.-H., Lin, C.-H., & Wong, C.-H. (2004). Studies on the immunomodulating and anti-tumor activities of *Ganoderma lucidum* (Reishi) polysaccharides. *Bioorganic and Medicinal Chemistry*, *12*(21), 5595–5601.

Chirivi, R. G., Garofalo, A., Crimmin, M. J., Bawden, L. J., Stoppacciaro, A., Brown, P. D., & Giavazzi, R. (1994). Inhibition of the metastatic spread and growth of B16-BL6 murine melanoma by a synthetic matrix metalloproteinase inhibitor. *International Journal of Cancer*, *58*, 460–464.

Davidson, B., Goldberg, I., Liokumovich, P., Kopolovic, J., Gotlieb, W. H., Lerner-Geva, L., Reder, I., Ben-Baruch, G., & Reich, R. (1998). Expression of metalloproteinases and their inhibitors in adenocarcinoma of the uterine cervix. *International Journal of Gynecology Pathology*, *17*, 295–301.

Donna,S., 2009. University of Michigan School of Medicine Chemical Reviews, 09, 7, 2859–2861.

el-Deiry, W. S., Harper, J. W., O'Connor, P. M., Velculescu, V. E., Canman, C. E., Jackman, J., Pietenpol, J. A., Burrell, M., Hill, D. E., & Wang, Y. (1994). WAF1/CIP1 is induced in p53-mediated G1 arrest and apoptosis. *Cancer Research*, *54*, 1169–1174.

Ferrara, N., & Henzel, W. J. (1989). Pituitary follicular cells secrete a novel heparin-binding growth factor specific for vascular endothelial cells. *Biochemical and Biophysical Research Communications*, *161*, 851–858.

Gnarra, J. R., Zhou, S., Merrill, M. J., Wagner, J. R., Krumm, A., Papavassiliou, E., Old- field, E. H., Klausner, R. D., & Linehan, W. M. (1996). Post-transcriptional regulation of vascular endothelial growth factor mRNA by the product of the VHL tumor suppressor gene. *Proceedings of the National Academy of Sciences of the United States of America*, *93*, 10589–10594.

Higginson, I. (Ed.) (1993). *Clinical audit in palliative care*. Oxford: Radcliffe Medical Press.

Hofseth, L. J., & Wargovich, M. J. (2007). Inflammation, cancer, and targets of ginseng. *The Journal of Nutrition*, *137*, 183–185.

Houck, J. C., Sharma, V. K., & Hayflick, L. (1971). Age differences in human skin collagen. *Cell and Tissue Culture Models in Dermatological Research*, *137*, 331–333.

Hsu, H. Y., Hua, K. F., Lin, C. C., Lin, C., Hsu, J., & Wong, C. (2004). Extract of Reishi polysaccharides induces cytokine expression via TLR4-modulated protein kinase signaling pathways. *Journal of Immunology*, *173*(10), 5989–5999.

Hutchison, A., & Glover, D. M. (1995). *Cell cycle control*. Oxford: IRL Press.

https://www.cancer.gov/about-cancer/causes-prevention/research

https://www.cancer.gov/about-cancer/treatment/types

https://www.mayoclinic.org/tests-procedures/cancer-treatment/about/pac-20393344

Kim, J., Li, B., Winer, J., Armanini, M., Gillett, N., Phillips, H. S., & Ferrara, N. (1993). Inhibition of vascular endothelial growth factor-induced angiogenesis suppresses tumour growth *in vivo*. *Nature*, *362*, 841–844.

Knox, J. D., Wolf, K., McDanie, C., Clark, V., Loriot, M., Bowden, G. T., & Nagle, R. B. (1996). Matrilysin expression in human prostate. *Molecular Carcinogenesis*, *15*, 57–63.

Kuttan, R., Bhanumathy, P., Nirmala, K., & George, M. C. (1985). Potential anticancer activity of turmeric (*Curcuma longa*). *Cancer Letters*, *29*, 197–202.

Lai, C. Y., Hung, J. T., Lin, H.-H., Yu, A. L., Chen, S.-H., Tsai, Y.-C., Shao, L.-E. Yang, W.-B., & Yu, J. (2010). Immunomodulatory and adjuvant activities of a polysaccharide extract of *Ganoderma lucidum in vivo* and *in vitro*. *Vaccine*, *28*(31), 4945–4954.

Lee, T. K., Johnke, R. M., Allison, R. R., Obrien, K. F., & Dobbs, L. J. (2005). Radioprotective potential of ginseng. *Mutagenesis*, *20*, 237–243.

Liu, D., & Chen, Z. (2009). The effects of cantharidin and cantharidin derivates on tumour cells. *Anti-Cancer Agents in Medicinal Chemistry*, *9*, 392–396.

Liu, L., Lassam, N. J., Slingerland, J. M., Bailey, D., Cole, D., Jenkins, R., & Hogg, D. (1995). Germline p16(INK4A) mutation and protein dysfunction in a family with inherited melanoma. *Oncogene*, *11*, 405–412.

Lu, C. X., Nan, K. J., & Lei, Y. (2008). Agents from amphibians with anticancer properties. *Anticancer Drugs*, *19*, 931–939.

Mantovani, A., Allavena, P., Sica, A., & Balkwill, F. (2008). Cancer related inflammation. *Nature*, *72*(3), 436–444.

Marx, J. (2004). Inflammation and cancer: The link grows stronger. *Science*, *306*(5698), 966–968.

Millauer, M. P., Longhi, K. H., Plate, L., Shawver, K., Risau, W., Ullrich, A., & Strawn, L. M. (1996). Dominant-negative inhibition of Flk-1 suppresses the growth of many tumor types *in vivo*. *Cancer Research*, *56*, 1615–1620.

Olofsson, B., Pajusola, K., Kaipainen, A., von Euler, G., Joukov, V., Saksela, O., Orpana, A., Pettersson, R. F., Alitalo, K., & Eriksson, U. (1996). Vascular endothelial growth factor B, a novel growth factor for endothelial cells. *Proceedings of the National Academy of Sciences of the United States of America*, *93*, 2576–2581.

Orlandini, M., Marconcini, L., Ferruzzi, R., & Oliviero, S. (1996). Identification of a c-fos-induced gene that is related to the platelet-derived growth factor/vascular endothelial growth factor family. *Proceedings of the National Academy of Sciences of the United States of America*, *93*, 11675–11680.

Ruddon, R. W. (1995). *Cancer biology* (3rd ed.). New York: Oxford University Press.
Seeger, R. C., Brodeur, G. M., Sather, H., Dalton, A., Siegel, S. E., Wong, K. Y., & Hammond, D. (1985). Association of multiple copies of the N-myc oncogene with rapid progression of neuroblastomas. *The New England Journal of Medicine*, *313*, 1111–1116.
Senger, D. R., Galli, S. J., Dvorak, A. M., Perruzzi, C. A., Harvey, V. S., & Dvorak, H. (1983). Tumor cells secrete a vascular permeability factor that promotes accumulation of ascites fluid. *Science*, *219*, 983–985.
Shukla, Y., & Kalra, N. (2007). Cancer chemoprevention with garlic and its constituents. *Cancer Letters*, *247*, 167–181.
Shu-Yi, Y., Wei, W.-C., Jian, F.-Y., & Yang, N.-S. (2013). Therapeutic applications of herbal medicines for cancer patients. *Hindawi Publishing Corporation*, *2013*, 1–15.
Singh, S. K. (2017). Cancer immunoprevention and public health. *Frontiers in Public Health*, *5*, 1.
Stats, H. F., Bradney, C. P., Gwinn, W. M., Jackson, S. S., Sempowski, G. D., Liao, H. X., Letvin, N. L., & Haynes, B. F. (2001). Cytokine requirements for induction of systemic and mucosal CTL after nasal immunization. *The Journal of Immunology*, *167*, 5386–5394.
Tessler, S., Rockwell, P., Hicklin, D., Cohen, T., Levi, B. Z., Witte, L., Lemischka, I., & Neufeld, G. (1994). Heparin modulates the interaction of VEGF165 with soluble and cell associated flk-1 receptors. *Journal of Biological Chemistry*, *269*, 12456–12461.
Ueki, N., Nakazato, M., Ohkawa, T., Ikeda, T., Amuro, Y., Hada, T., & Higashino, K. (1992). Excessive production of transforming growth-factor beta 1 can play an important role in the development of tumorigenesis by its action for angiogenesis: validity of neutralizing antibodies to block tumor growth. *Biochimica et Biophysica Acta*, *1137* 189–196.
Wang, G. S. (1989). Medical uses of *Mylabris* in ancient China and recent studies. *Journal of Ethnopharmacology*, *26*, 147–162.
Wang, Y., Qian, X. J., Hadley, H. R., & Lau, B. H. (1992). Phytochemicals potentiate interleukin-2 generated lymphokine-activated killer cell cytotoxicity against murine renal cell carcinoma. *Molecular Biotherapy*, *4*, 143–146.
Weidner, N., Semple, J. P., Welch, W. R., & Folkman, J. (1991). Tumor angiogenesis and metastasis: Correlation in invasive breast carcinoma. *The New England Journal of Medicine*, *324*, 1–8.
Witty, J. P., McDonnell, S., Newell, K. J., Cannon, P., Navre, M., Tressler, R. J., & Matrisian, L. M. (1994). Modulation of matrilysin levels in colon carcinoma cell lines affects tumorigenicity *in vivo*. *Cancer Research*, *54*, 4805–4812.
World Health Organization. (1986). *Cancer pain relief and palliative care.* WHO.
World Health Organization. (1996). *Cancer pain relief and palliative care.* WHO.
World Health Organization. (1998). *Symptom relief in terminal illness.* WHO.
World Health Organization. (2002). *National cancer control programmes: Policies and management guidelines* (2nd ed.). WHO.
Yoshida, Y., Wang, M. Q., Liu, J. N., Shan, B. E., & Yamashita, U. (1997). Immunomodulating activity of Chinese medicinal herbs and *Oldenlandia diffusa* in particular. *International Journal of Immunopharmacology*, *19*, 359–370.
Zafonte, T., Hulit, J., Amanatullah, D. F., Albanese, C., Wang, C., Rosen, E., Reutens, A., Sparano, J. A., Lisanti, M., & Pestell, R. G. (2000). Cell-cycle dysregulation in breast cancer: breast cancer therapies targeting the cell cycle. *Frontiers in Bioscience*, *5*, D938–D961.

2 Traditional System of Medicines

Plant-derived Compounds for Cancer Therapy

Ankur Joshi, Varsha Johariya, Neelesh Malviya, and Sapna Malviya

CONTENTS

2.1 CURRENT SCENARIO OF TRADITIONAL MEDICINES

Traditional medicines (TMs) are used by a considerable portion of the population in developing nations, either because of the exorbitant cost of Western medications and healthcare or because traditional remedies are more acceptable from a cultural and spiritual standpoint. According to World Health Organization (WHO), 80% of

DOI: 10.1201/9781003251712-2

people in developing countries still rely on herbal medicines for healthcare. India is the richest source of herbals, with approximately 25,000 plant-based formulations being used by all communities, including rural and folk. India shares 0.2% export of herbals, that is US$63 billion. The need of the hour is to focus on standardization of these herbal traditional products. This will indeed improve quality as well as its proper use. So pharmacovigilance of herbals can help majorly in this direction to detect, assess, understand, and prevent the adverse effects related to herbal, traditional, and complementary medicines.

Cancer is a disease that affects the human population worldwide. New medicines to treat and prevent this life-threatening condition are always in demand. Natural secondary metabolites (SMs) produced by plants are being studied for their anti-cancer properties, which could lead to the creation of novel therapeutic medications. As a result of the success of these substances, which have been turned into standard cancer treatments, new technologies are emerging to further develop the field. Nanoparticles for nanomedicines are a novel technology that aims to improve the anti-cancer properties of plant-derived medications by controlling the release of the molecule and testing new delivery methods. The demand for naturally produced chemicals from medicinal plants, as well as their features that make them viable anti-cancer treatments, is discussed in this chapter.

2.1.1 Integrated Approaches for Development of Herbal Medicine

The majority of pharmaceutical businesses, including multinationals, have become interested in the international trade in herbal medicine. A few years ago, only small businesses were interested in selling herbal medications. Several significant international corporations are currently engaged in commercializing herbal medicines. The global herbal medicine market, which includes herbal products and raw materials, is expected to develop at a rate of up to 15% each year. Figure 2.1 depicts several integrated herbal research methodologies for the promotion and development of natural products.

2.1.2 Plant Constituents with Anti-cancer Properties

Medicinal plants have been utilized in TM in Asian and African populations for thousands of years, and many plants in modern countries are taken for their health advantages. According to WHO, some countries still rely on plant-based medicine as their primary supply of medicine, while developing countries are taking advantage of the therapeutic benefits of organically produced chemicals. Polyphenols, brassinosteroids (BRs), and Taxols are examples of anti-cancer compounds that have been found and isolated from terrestrial plants.

2.1.2.1 Polyphenols

Polyphenolic compounds such as flavonoids, tannins, curcumin, resveratrol, and gallocatechin have anti-cancer properties. Various foods contain resveratrol, including peanuts, grapes, and red wine. Green tea contains gallocatechin, which is antioxidant. Polyphenols, which are natural antioxidants, are thought to boost health and reduce cancer risk when included in a person's diet. They induce apoptosis and so

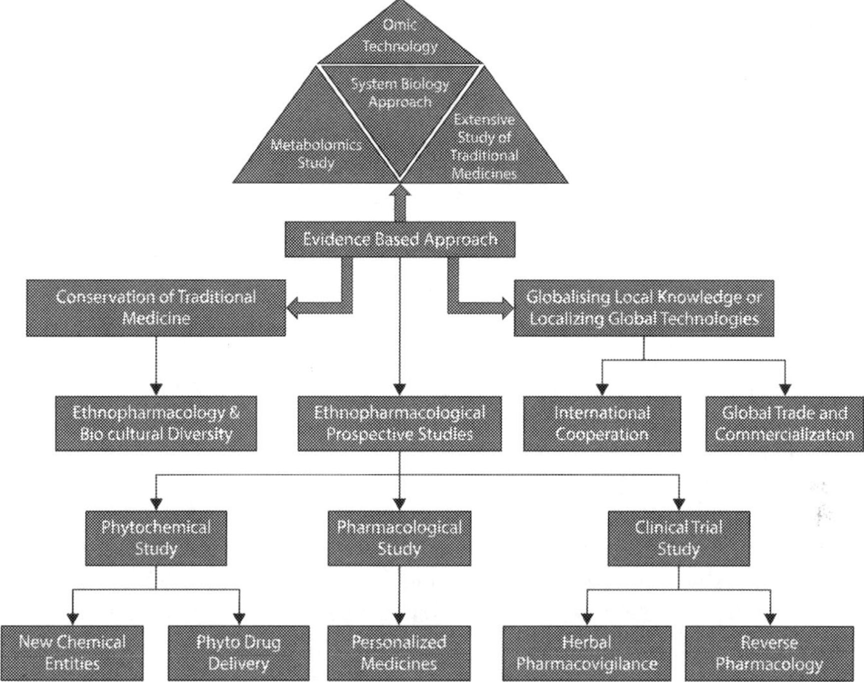

FIGURE 2.1 Integrated approaches for development of herbal medicine.

have anti-cancer qualities that can be used and they are considered to initiate apoptosis by controlling the mobilization of copper ions linked to chromatin, causing DNA breakage. Polyphenols cytotoxicity has been established in a variety of cancer cells, and their antioxidant properties have been determined. Plant polyphenols also have the ability to inhibit the growth of cancer cells by interfering with proteins found in cancer cells. By directly interacting with cancer agents, the polyphenol can regulate acetylation, methylation, or phosphorylation. Resveratrol was found to be capable of DNA destruction in the presence of Cu(II). Curcumin-treated cancer cells in a variety of cell lines, for example, have exhibited inhibition of tumour necrosis factor (TNF) expression in response to various stimuli.

2.1.2.2 Flavonoids

More than 10,000 different flavonoids have been found in plants, making them one of the most diverse groups of SMs in nature. Biologically active molecules in plants that are being studied for their potential health benefits are known as phytonutrients. Traditional Chinese medicine plants like litchi leaf and fern species have flavonoid content investigated to see if these substances affect cancer cells. Anthocyanin, flavones, flavonols, chalcones, and other flavonoid compounds are concentrated in the seed of the plant. Flavonoids from the *Dryopteris erythrosora* fern were tested on human lung cancer cells by Coa et al. (2013) (A456 cell line). It was found that flavonoids can kill and scavenge free radicals in cancer cells, as well as cytotoxicity.

Additionally, purified flavonoids have been shown to have anti-cancer effects against hepatoma, cervical cancer, and breast cancer in humans (MCF-7). HL-60 cells were used to test the cytotoxicity of flavonoids extracted from *Erythrina suberosa* stem bark (4′-methoxy licoflavanone [MLF] and alpinumi [AIF]) (human leukaemia). Apoptosis was induced by MLF and AIF via both internal and external signalling mechanisms. The mitochondrial membrane potential is greatly reduced when apoptotic proteins are produced. Injuries to the mitochondria of cancer cells make them vulnerable to death. Flavonoid extracts from fern species have been studied and found to have a substantial anti-cancer effect even at low levels.

It has previously been reported that polyphenols can influence the control of proteins and other agents that may contribute to the survival of cancer cells. Proliferation of cancer cells is facilitated by STAT (signal transducer and activator of transcription) proteins, which are anti-apoptotic. MLF and AIF suppress the phosphorylation of members of this family of proteins, which is essential for cancer cell survival. These flavonoids also inhibit NF-κB, which is essential for cancer cell survival, angiogenesis, and proliferation.

2.1.2.3 Brassinosteroids

BRs are naturally occurring substances found in plants that govern cell development and differentiation, as well as elongation of stem and root cells and other functions such as disease and stress resistance and tolerance. BRs are also employed to control plant senescence. They are necessary for the growth and development of plants. BRs are naturally occurring chemicals that have shown therapeutic value in the fight against cancer.

Two natural BRs have been employed in studies involving malignant cells to show that they have anti-cancer capabilities. The anti-cancer compounds 28-homocastasterone (28-homoCS) and 24-epibrassinolide (24-epiBL) have been shown to be effective at micromolar doses in a variety of cancer cell lines. Cancer cells are unique in that they do not naturally undergo apoptosis and continue to grow continuously. By interacting with the cell cycle, BRs can trigger reactions that restrict growth and cause apoptosis. T-lymphoblastic leukaemia (CEM), multiple myeloma, cervical cancer (HeLa), lung carcinoma A-549, and osteosarcoma (HOS) cell lines have all been treated with this medication. Breast and prostate cancer cell lines are also treated. Treatments for breast cancer target proteins such as the oestrogen and epidermal growth factor receptors (HER2) and the epidermal growth factor receptor (EGFR) because of their high abundance in breast cancer cells such as MCF-7 and MDA-MB-468. It is important to note that the androgen receptor (AR) is similar in structure to the oestrogen receptor (OR), which has a role in the proliferation of prostate cancer cells (ER). BRs will interact with or bind to these proteins' receptors, inhibiting the proliferation of hormone-sensitive and hormone-insensitive cancer cells alike. BRs can also cause cell cycle disruption. Cyclin proteins involved in the G1 cell cycle phase were reduced in breast cancer cell lines treated with 28-homoCS and 24-epiBL. Cells will either repair or enter apoptosis at this point in the cell cycle; BRs encourage apoptosis at this stage, which cancer cells would not ordinarily be able to do. The balance of apoptotic proteins that promote cell survival versus those that induce programmed cell death shifts in prostate cancer cell lines LNCaP and

DU-145 after BRs treatment. After BRs therapy, levels of the Bax pro-apoptotic protein rise, whereas anti-apoptotic proteins like Bcl-2 fall. BRs cause diverse reactions in normal and malignant cells, in addition to their anti-cancer capabilities. Agents of BRs origin are of interest for therapeutic qualities since they are not cytotoxic to normal cells and are cell-specific to cancer cells, which is a fundamental requirement in anti-cancer treatment.

2.2 SCIENTIFIC VALIDATION OF TRADITIONAL MEDICINAL PLANTS

Various critical areas in medicinal plant research are being considered to reinvigorate herbal medicine and bring it in line with modern medicine. Scientists believe that combining herbal medicine with contemporary instruments will not only improve their own development but will also aid in the treatment of a variety of difficult ailments through the creation of new entities. It is true that a substantial number of medicines are produced from plants or synthetic counterparts developed from plants. Only with sophisticated methodologies and unique strategies will dedicated research be effective. There are numerous methods for determining the quality of natural and manufactured substances. Traditional drug discovery methodologies presently use a variety of *in vitro*, *in vivo*, and high-throughput screening technologies.

Interest in natural remedies, particularly in herbal therapy, among people has grown rapidly in recent decades, not only in developing countries but also in developed countries. This has greatly enhanced international herbal medicine trade and attracted a majority of pharmaceutical enterprises, including multinationals. India is one of the few countries capable of growing a majority of the key plants utilized in both modern and traditional medical systems. Combinatorial chemistry and high-throughput screening are very effective approaches in the present day, and herbal resources are producing a lot of new therapeutic compounds. From the standpoint of quality, safety, and efficacy, the traditional usage of medicinal herbs needs to be rigorously examined and standardized. Although there has been a surge in interest in science-based herbal medicine research, most of it has been hampered by studies that have used unauthenticated, uncharacterized materials. The quality of the test material is one of the most significant aspects of any research project. If the material studied was not validated and characterized in such a way that it could be duplicated, a study cannot be regarded scientifically genuine. In the case of botanicals, misidentification of the harvested plant, adulteration with other species, or contamination with foreign substances are all possible.

2.3 TRADITIONAL MEDICINES INSPIRED DRUG DISCOVERY AND DRUG DEVELOPMENT

TM is beneficial in all phases of natural resource drug development. A few instances of pharmaceuticals made from natural sources would help to explain the tradition's history. For many years, scientists had used several TM-based techniques for drug discovery and development. Artemisinin (antimalarial), vincristine, vinblastine, camptothecin podophyllotoxin, etoposide, teniposide, and paclitaxel (anti-cancer) are only a few of the therapeutically promising compounds discovered from plants. Pharmaceutical companies are doing a large-scale pharmacologic screening of herbs in order to produce medications from ayurvedic botanicals. The plant *Commiphora mukul* Hook was mentioned in *Sushrutae Samhita*, a Sanskrit classic on Ayurveda written about 600 BC, as being effective in the treatment of obesity and related disorders. Several therapeutically powerful leading molecules from TM may be influenced by a confluence of exceptional advancements in chemistry, molecular biology, genomics, and chemical technology, as well as the cognate domains of spectroscopy, chromatography, and crystallography, in recent years.

2.4 SECONDARY METABOLITES IN CANCER TREATMENT

Plants produce a vast range of SMs, which act as both defence and signal molecules against herbivores, other plants, and microorganisms. SM has a diverse range of biological and pharmacological characteristics. As a result, some of the plants or compounds isolated from them have been and continue to be utilized to treat infections, health problems, and diseases.

Plants have long been a rich source of inexpensive natural chemicals, particularly SMs, with sufficient structural complexity to make their synthesis difficult or impossible at the present time, and a wide range of bioactivities, including anti-cancer activity. SMs are tiny organic molecules produced by an organism but not required for its growth, development, or reproduction. The pathway via which they are generated can be used to classify them: terpenoids (polymeric isoprene derivatives biosynthesized via the mevalonic acid pathway), phenolics (biosynthesized via shikimate pathways, having one or more hydroxylated aromatic rings), and alkaloids (very varied) are divided into three categories (non-protein nitrogen-containing compounds, biosynthesized from amino acids such as tyrosine, with a long history in medication). In the fight against cancer, new and potentially dangerous SMs are isolated from plants on a yearly basis.

Because of their physicochemical features (e.g. restricted bioavailability) and/or toxicity, certain natural substances with unique anti-cancer activities cannot be used in clinical practice set. SMs found in plants, on the other hand, can frequently be good drug development leads. Modifying the chemical structure of these more promising compounds is thus one strategic strategy to improve their anti-cancer efficacy and selectivity, as well as their absorption, distribution, metabolism, and excretion capabilities while lowering their toxicity and side effects. The most significant breakthroughs in the field of plant SMs, some of which are now in clinical use or are in clinical trials as anti-cancer medicines, as well as their most effective derivatives created by structural alterations, are mentioned in this chapter.

2.5 GENERAL MECHANISM OF ACTION OF MEDICINAL PLANTS

Alkaloids, flavonoids, anthraquinones, saponins, glycosides, carbohydrates, tannins, antioxidants, proteins, polyphenols, and steroids are some of the active phytoconstituents found in medicinal plants for the management of cancer.

2.5.1 Alkaloids

Apoptosis serves a critical function in multicellular organisms' development and survival. DNA breakage/laddering, cell shrinkage, nuclear condensation, membrane blebbing (membrane-bound apoptotic entities), and regulation of precise signalling circuitry are all examples of well-executed morphological and biochemical phenomena that precede cell death. As a result, apoptosis is an important aspect of the regular cell cycle and a natural defence against cancer. Tumourigenesis is caused by irreversible changes in the apoptotic machinery's upstream regulators or downstream effector components. In fact, one of the ten characteristics of cancer, escaping apoptosis and opposing cell death, is what drives the process of cell immortalization. Nonetheless, any drug capable of inducing and activating apoptotic signalling proteins can be regarded as a prospective cancer therapeutic option.

- *DNA Damaging Alkaloids:* To repair their most vulnerable substance, DNA, cells have evolved two ways. The first is the immediate repair of damaged DNA, and the second is the inactivation of cells with damaged genomes. Both procedures are necessary to maintain cellular genomic stability since unrepaired DNA damage is frequently associated with genetic mutations, which can lead to malignant transformation.
- *Apoptotic Alkaloids: Caspase Activators:* In a self-amplifying cascade, cysteine aspartic proteases, or "caspases," preferentially break their substrate proteins at the aspartate residues during apoptosis. Caspases play a vital role in the death of cells that have been programmed to die. The intrinsic and extrinsic mechanisms have been identified as two primary caspase activation pathways. They combine DNA damage, cytotoxic chemicals, ROS generation, abnormal oncogene expression, and p53 activation to activate multiple apoptotic signals.
- *Anti-proliferative Alkaloids—Cell Growth Inhibitors:*
  1. *Cell Cycle Arrest:* DNA replication and cellular division are processes that rely on the coordination of many protein phosphorylation cascades, as well as on a number of checkpoints that oversee these two key activities and ultimately lead to cell cycle completion. Cell cycle deregulation is one of the ten hallmarks of cancer transformation, and targeting one of its primary modulators, such as cyclins, cyclin-dependent kinases (CDKs), and tumour suppressor proteins (p53 and Rb), is an important step in stopping cell cycle and causing cancer cell death.
  2. *Alteration of the MAPK Pathway:* The MAPK pathway links extracellular signals to cellular responses and processes like growth, proliferation, differentiation, migration, and death in normal cells. A specific ligand interacts to a tyrosine receptor, phosphorylating the Ras protein and activating it. Thereafter, the latter attaches to effector proteins such as B-Raf, which activate MEK 1 and MEK 2. These kinases activate ERK 1 and ERK 2, which in turn activate a number of AP-1 transcription factors. These agents eventually make their way to the nucleus, where they trigger the production of genes that code for growth factors, cyclins, cytokines, and other

proteins involved in cell proliferation. To avoid overexpression of the listed proteins, the Ras-GTP-B-Raf complex is normally inactivated quickly after activation by GTPase-activating protein (GAP).

3. *Suppression of the NF-κB Pathway:* Nuclear factor-kappa B (NF-κB) is an inducible nuclear transcription factor that regulates cell survival and proliferation by activating genes. Downregulating it, on the other hand, causes the cells to go into apoptosis. The upregulation of proteins involved in cell invasion and angiogenesis, particularly those linked to cell adhesion and survival, occurs when this pathway is activated abnormally. Alkaloids have been discovered to inhibit carcinogenesis by decreasing the NF-κB pathway's activation and modulating its gene expression.

- *Other Deadly Mechanisms—An Infinite Diversity Promyelocytic:* Many of the cited developing alkaloids have been shown to function in a variety of molecular ways, as shown throughout this chapter. Additional modes of action were discovered in the hirsutine and carboline alkaloids, which are discussed below.
- *Formation of G-Quadruplexes:* G-quadruplexes, which are made up of four guanine bases arranged in a three-dimensional square planar structure, is formed by carboline alkaloids. These arrangements operate as anti-cancer agents against human promyelocytic leukaemia, prostate cancer, and gastric cancer by regulating genes, particularly oncogenes. The production of G-quadruplexes can be predicted as a novel cancer therapy strategy with potential implications in anti-cancer medication development.
- *HER2 Targeting:* The most successful anti-cancer medications are those that target certain types of chemicals. Because it selectively targets the HER2 proteins activated in breast cancer cells, the discovery of the alkaloid hirsutine piqued the curiosity of numerous researchers. In reality, the ERBB2 oncogene, which is produced in normal breast cells and leads to the insertion of HER2 receptors, encodes these proteins. These molecules act as a control system for the cell's growth, division, and repair mechanisms. A mutation in ERBB2 causes gene amplification, resulting in the synthesis of a large number of HER2 receptors, or a change that causes the HER2 protein to be overexpressed, both of which can lead to a loss of control over cell division and overgrowth. Breast cancer cells that are HER2-positive grow quicker, spread faster, and have a larger likelihood of relapse than normal breast cancer cells, making them more dangerous and life-threatening. As a result, anti-HER2 cancer drugs like hirsutine work by interfering with cancer cells' excessive growth impulses, causing them to die. The MDA-MB-453 and BT474 cell lines, for example, showed cytotoxic reactions and apoptosis after being treated to hirsutine.
- *Inhibition of the P-Glycoprotein ABCB1:* It is a novel alkaloid with an unusual molecular mode of action on various cancer cells. In reality, by inhibiting P-glycoprotein (P-gp), also known as the ABCB1 member of the ABC proteins, this metabolite isolated from the flowering plants Amaryllidaceae fights breast cancer, cervical cancer, and skin epidermoid

carcinoma. Multidrug resistance is a severe problem in cancer treatment, and it can be caused by the upregulation or activation of ATP-binding cassette (ABC) proteins, which make cancer cells resistant to therapies. As a result, inhibiting P-gp, one of the most essential ABC proteins, improves the efficacy of cancer-targeting cytotoxic medicines.

2.5.2 Flavonoids

Polyphenolic substances have a wide range of biological actions, including those that may influence the dysregulated processes that occur during cancer formation. Antiallergic, anti-inflammatory, antioxidant, antimutagenic, anticarcinogenic, and enzymatic activity modulation are all examples. As a result, they may have favourable health effects and can be used as cancer chemopreventive or therapeutic agents.

2.5.2.1 Major Molecular Mechanism of Action

Carcinogenesis is a complicated and multistep process in which unique molecular and cellular modifications occur, and three stages have been defined to simplify the various choices for chemoprevention and chemotherapy in cancer development and progression:

- Initiation is a quick step that involves exposing cells to a carcinogenic agent and allowing them to interact with it, particularly DNA.
- Because promotion lasts longer than the previous stage, aberrant cells can persist, multiply, and form a cluster of preneoplastic cells.
- The progression stage of carcinogenesis includes the steady conversion of premalignant cells to neoplastic cells, as well as an increase in invasiveness, metastatic potential, and the development of new blood vessels (angiogenesis).

The discovery of oncogenes is one of the most exciting discoveries. Over 40 oncogenes have been discovered, and their protein products have been studied. Protein kinases, GTP-binding proteins, and nuclear transcription factors are among them. It was proposed that the initiation of transformation might be mediated by protein phosphorylation. Protein-tyrosine kinases (PTKs) are a family of enzymes that catalyse the transfer of ATP's phosphate to the hydroxyl group of tyrosine on a variety of important proteins, resulting in a cascade of altered cell parameters, which is a hallmark of transformed cells. Several recent discoveries have backed up this hypothesis:

1. Activated PTKs are the products of almost half of all known viral transforming genes (oncogenes).
2. Among the plasma membrane receptors for ligand-activated PTKs are those for epidermal growth factor (EGF), transforming growth factor (TGF), platelet-derived growth factor (PDGF), insulino-like growth factor-1 (IGF-1l), macrophage colony-stimulating factor-1 (CSF-1l), fibroblast growth factors (FGF-1 and FGF-2), nerve growth factor (NGF), and hepatocyte growth factor (HGF).

2.5.2.2 Preventing Carcinogens Metabolic Activation

Phase of activation—one of the most important routes by which flavonoids might exert their effects—occurs by metabolizing enzymes (e.g., cytochrome P450). These enzymes metabolically activate a vast variety of procarcinogens, resulting in the reactivation of intermediates that can interact with cellular nucleophiles and, eventually, cause carcinogenesis. Several P450 isozymes, including CYP1A1 and CYP1A2, are considered to be inhibited by flavonoids, and as a result, cells are believed to be protected against the cellular damage induced by carcinogen activation. Another mechanism of action is the stimulation of phase II metabolizing enzymes (e.g., GST, quinone reductase, and UDP-GT), which detoxify and eliminate carcinogens from the body. This contributes to the understanding of flavonoids' anti-carcinogenic and chemopreventive effects.

2.5.2.3 Anti-proliferation

The suppression of the pro-oxidant process that promotes tumour growth could be a molecular mechanism of anti-proliferation. The main catalysts of tumour development and progression are growth-promoting oxidants and reactive oxygen species (ROS). Flavonoids suppress the enzyme xanthine oxidase, also known as COX or LOX55, and thereby tumour cell proliferation.

Furthermore, the mechanism of prevention of polyamine production may contribute to flavonoids' anti-proliferative properties. Ornithine decarboxylase is a rate-limiting enzyme in polyamine biosynthesis, and its activity is linked to DNA synthesis and cell proliferation in a variety of tissues. Flavonoids have been shown in several studies to block ornithine decarboxylase, which is stimulated by tumour promoters, resulting in a decrease in polyamine and suppression of DNA and protein synthesis.

2.5.2.4 Cell Cycle Arrest

Flavonoids' anticarcinogenic effects may be due to changes in cell cycle progression. Cells undertake a series of regulated phases permitting cell cycle progression in response to mitogenic signals, and CDKs are recognized as important regulators of cell cycle progression. CDK activity alteration and dysregulation are pathogenic markers of neoplasia. CDK hyperactivation has been linked to a variety of malignancies owing to mutations in CDK genes or CDK inhibitor genes. As a result, inhibitors or modulators are attracting a lot of attention as potential cancer therapeutics. Flavonoids such as silymarin, genistein, quercetin, daidzein, luteolin, kaempferol, apigenin, and epigallocatechin 3-gallate have been shown in cultured cancer cell lines to disrupt checkpoints at both the G1/S and G2/M stages of the cell cycle. According to research from numerous laboratories, flavopiridol inhibits all CDKs, producing cell cycle arrest in either G1 or G2/M.

2.5.2.5 Induction of Apoptosis

Flavonoids' potent anti-cancer activities could be attributed to frank apoptosis. Apoptosis is a type of active cell death that aids in the development and survival of organisms by removing damaged or undesirable cells. It is closely regulated by a collection of genes that support apoptotic cell survival, and it is mediated by a highly structured network of interacting proteases and inhibitors in response to unpleasant stimuli from both inside and outside the cell. Apoptosis dysregulation is a key factor in cancer development. Flavonoids have been demonstrated to cause apoptosis in cancer cell types while leaving healthy cells alone. The molecular mechanisms through which flavonoids cause apoptosis are unknown. Mechanisms that may be involved in the inhibition of DNA topoisomerase I/II activity include reductions in ROS, manipulation of signalling pathways, downregulation of nuclear transcription factor kappa B (NF-κB), and endonuclease activation and suppression of Mcl-1 protein.

2.5.2.6 Flavonoids Preventing Carcinogens Metabolic Activation

In vitro investigations on the possible anti-cancer effect of flavonoids in various cell systems have been undertaken by several researchers. Hirano and colleagues compared the anti-cancer efficacy of 28 flavonoids on the human acute myeloid leukaemia cell line HL-60, as well as the discrepancies between their anti-proliferative activity and cytotoxicity to that of four therapeutic anti-cancer drugs. Eight of the 28 flavonoids examined inhibited HL-60 cell proliferation significantly, with IC_{50} values in the range of 10–940 ng/ml. Honokiol, machilin A, matairesinol, and arctigenin exhibited the greatest effects, with IC_{50} values of less than 100 ng/ml, which were substantially equivalent to the effects of previously recognized anti-cancer drugs. For HL-60 cells ($LC_{50} > 2{,}900$ ng/ml), dye exclusion studies found flavonoids such as genistein and lignans to be non-cytotoxic ($LC_{50} = 2{,}900$ ng/ml), whereas normal anti-cancer drugs were very toxic to these cells. More than 30 flavonoids were evaluated in human colon cancer cell lines Caco-2 and HT-29 for their impact on cell proliferation and probable cytotoxicity.

2.5.3 Anthraquinones

The anthraquinones core has a stiff structure and is made up of a planar three-ring aromatic anthracene system with two keto groups at the 9- and 10-positions. The role of anthraquinones in the DNA double helix of cancer cells is to conduct a specific redox cycle *in vivo*, generating the superoxide radical anion (O^{2}–). Natural anthracene-9,10-diones, such as emodin, aloe emodin, and rhein, have often been exploited as a starting point in the development of anti-cancer medicines throughout the last few decades. Many anthraquinone derivatives have been discovered and manufactured, and many medications, such as mitoxantrone, doxorubicin, and Epirubicin, have already been licensed for clinical use. Although anthraquinone compounds have been shown to have some cardiotoxicity in clinical trials, recent research has found that forming anthraquinone complexes with metals (Cu, Co, Fe, and Mn) can significantly reduce the formation of ROS and thus their toxic and side effects. Anthraquinones molecules with strong targeting can also be achieved through improved drug delivery systems, reducing harm to normal cells and tissues. As a result, research on novel anthraquinones with low toxicity, excellent targeting, and good drug ability is still worthwhile. Previous research has discovered that the anti-cancer mechanism of anthraquinone chemicals can be separated into three subtypes: DNA damage, cell cycle arrest, and apoptosis. Furthermore, these drugs have been shown to have anti-cancer effect by different pathways such as paraptosis, autophagy, abnormal cellular metabolism, radio sensitization, anti-invasion, and anti-metastasis. The recently discovered novel anthraquinone derivatives, in particular, have the potential to target the physiological impacts of altered critical signalling pathways.

- *Anthraquinone Anti-cancer Agents Targeting DNA:* DNA is responsible for genetic information as well as the storage and transmission of information in living organisms' molecular processes. Furthermore, many cytotoxic anti-cancer drugs target it as a common cellular target. As a result, the development of anti-cancer medicines that target DNA is currently a hot topic. Several anthraquinone chemicals, including small-molecule derivatives and organometallic complexes, have been shown to be effective in the treatment of cancer by targeting DNA.
- *Anthraquinone Anti-cancer Agents Targeting Apoptosis:* Apoptosis is a natural physiological process that can be triggered by a variety of external causes. The signal transmission system in the apoptosis process, however, is still unknown. A membrane receptor pathway, such as Fas–FasL, and a biochemical mechanism involving cytochrome c and caspase activation are currently the two main obvious pathways. Bax and Bcl-2 are members of the Bcl-2 family, which regulates cytochrome c release in opposite ways during mitochondria-dependent apoptosis. Even though anthraquinone-induced apoptosis mostly occurs as described above, new anti-cancer anthraquinone derivatives are still needed.
- *Anthraquinone Anti-cancer Agents Inducing Paraptosis:* The molecular mechanism of paraptosis, a new non-caspase-dependent programmed cell death, is unknown. It appears to be distinct from apoptosis, autophagy necrosis, and other similar processes. Drug treatment failures are prevalent in tumour cells with an aberrant apoptotic pathway, and paraptosis may be a technique to destroy apoptotic cells that have escaped death, presenting a novel cancer chemotherapy strategy. Tian and colleagues have discovered that anthraquinone chemicals can cause cancer cell death via paraptosis. Paraptosis is produced by 1,8-bis(benzyloxy)-9,10-anthraquinone-*N*-(2-hydroxyethyl)-3-carboxamide in various malignant hepatocellular carcinoma cell lines, including SMMC-7721 and HepG2 cells, as well as human nasopharyngeal carcinoma (NPC) cells, including CNE1 and CNE2.
- *Anthraquinone Anti-cancer Agents Targeting Autophagy:* Although autophagy is a sort of programmed cell death, its role in cancer is complicated because it can produce tumours, increase cancer cell survival, and inhibit necrosis and even cell death. When cancer cells are exposed to hypoxia and malnutrition in their microenvironment, they are more likely to survive cell death through an autophagy pathway, which helps cancer cells adapt to changing conditions or prevents apoptosis and death. Autophagy plays two roles during carcinogenesis and tumour growth. As a result, developing and synthesizing new chemicals that activate cell autophagy will aid in furthering our understanding of the relationship between cell autophagy and tumours and, ultimately, conquering this disease. Researchers have discovered that a group of anthraquinone chemicals can cause cancer cells to die by inducing autophagy.

- *Anthraquinone Anti-cancer Agents in Sensitizers for Photodynamic Therapy:* Photodynamic treatment (PDT) is a novel therapeutic technique that combines a photosensitizer and a light source to use photodynamic reactions to selectively eliminate tumour tissues. PDT has become one of the most modern precision cancer treatment procedures due to its great selectivity and low invasiveness. In PDT, photosensitizers play a crucial role. They are exceedingly unstable when stimulated and return to the ground state in a very short period, resulting in a significant amount of ROS. The produced ROS subsequently interact with biological macromolecules in the human body, resulting in cell injury or changes in cell function, delivering the therapeutic goal.
- *Anthraquinone Anti-cancer Agents Induce Abnormal Cellular Metabolism:* Cancer cells require a lot of nutrients to grow quickly, and their metabolism is different from that of normal cells. Cancer cells require more glucose, although oxidative phosphorylation is rarely used by them. Given that glycolysis provides roughly 80% of cancer cell energy, glucose metabolism in cancer cells tends to be glycolytic. PGAM1 is an enzyme that converts 3-phosphoglycerate to 2-phosphoglycerate in the glycolysis process. Through the control of its substrate 3-phosphoglycerate and the equivalent product 2-phosphoglycerate, PGAM1 coordinates glycolysis, the pentose phosphate pathway, and serine production to promote tumour growth.
- *Anthraquinone Anti-cancer Agents Targeting Hypoxia:* When the partial pressure of oxygen in tumours is lower than the necessary level, hypoxia refers to specific biological and molecular repercussions. It is one of the inherent properties of most solid tumours and it can increase tumour cell tolerance to radiation and chemotherapy, as well as promote tumour cell angiogenesis and metastasis, while also serving as a significant negative prognostic factor. As a result, hypoxia therapy has received more attention, and the targeted killing of hypoxic cells in tumours has become a hot topic in cancer therapy. Anthraquinones are known to go through a redox cycle that produces the superoxide radical anion O^{2-}. The creation of O^{2-} sets in motion a metal-assisted cascade that results in the formation of H_2O_2 and hydroxyl radicals (OH^-). These produced radicals have the potential to harm cells by damaging unsaturated lipids, proteins, and DNA.
- *Chemotherapy and Radiotherapy:* Non-lethal touch causes resistance, which is a serious issue in cancer treatment. Although the exact process is still being investigated, the discovery and use of modified anthraquinone derivatives have the potential to overcome tumour cells' radio and drug resistance.

2.5.4 Saponins

Saponins are found in the roots, tubers, leaves, flowers, and seeds of a wide range of higher plants. Saponins were divided into triterpenes and steroids based on their carbon skeletons. Their glycone components were largely oligosaccharides, which were organized in a linear or branched pattern and linked to hydroxyl groups via an acetal linkage. Saponins have an anti-cancer effect on numerous cancer cells, according to recent studies. With IC_{50} values as low as 0.2 mm, some saponins suppress tumour cell growth by cell cycle arrest and death. Meanwhile, saponins, when used in conjunction with traditional tumour therapy procedures, improve therapeutic outcomes.

- *Mechanisms of the Anti-tumour Effect of Saponins:* An ERK-independent NF-κB signalling pathway has been shown to be modulated by several specific saponins with high anti-cancer effects. Furthermore, Astragalus saponins (AS) could be utilized as an adjuvant in conjunction with other standard chemotherapy medicines to mitigate the latter's negative effects. It would target the NSAID-activated gene (NAG-1) to minimize the synergistic effects of PI3K-Akt inhibitors when administered together. The findings could aid in the creation of a novel target-specific chemotherapeutic drug with well-understood molecular pathways in the future.
- *Cycloartanes:* Cycloartane saponins had a minor anti-cancer activity, but they could be employed as a cancer chemotherapeutic drug. Total AS, for example, are anti-cancer in human colon cancer cells and tumour xenografts. By triggering apoptosis and regulating an ERK-independent NF-κB signalling pathway, they reduced the production of the HCC tumour marker fetoprotein and decreased the HepG2 cell proliferation. Furthermore, AS could be utilized as an adjuvant in conjunction with other standard chemotherapy medicines to mitigate the latter's negative effects. It would target the NAG-1 to minimize the synergistic effects of PI3K-Akt inhibitors when administered together. The findings could aid in the creation of a novel target-specific chemotherapeutic drug with well-understood molecular pathways in the future.
- *Dammaranes:* The majority of dammarane saponins had anti-cancer properties. OSW-1, a naturally occurring chemical discovered in *Ornithogalum saudersiae* bulbs, is highly cytotoxic to tumour cell types. In contrast to cancer cells, non-malignant cells were statistically less susceptible to OSW-1, with dosages inducing a 50% loss of cell viability 40–150 times higher than those reported in cancer cells. According to electron microscopy and biochemical analyses, when the mitochondrial membrane and cristae were disrupted by OSW-1, apoptosis was initiated in both leukaemia and pancreatic cancer cells.
- *Oleananes:* Oleananes have the greatest saponins of any plant. Anti-tumour effects were achieved by a variety of mechanisms, including anti-cancer, anti-metastasis, immunostimulant, chemoprevention, and so on. Through distinct signalling transductions, avicins, tubeimoside, saikosaponins, platycodigenins, soybean saponin, and *Pulsatilla koreana* saponins showed anti-cancer effects.

2.6 CLINICALLY INVESTIGATED MEDICINAL PLANTS FOR THE TREATMENT OF CANCER

A wide range of anti-cancer properties have been discovered in numerous herbal medicines, according to a number of clinical investigations. We've categorized and classified the clinical use of a variety of herbal medications in this section based on their ability to inhibit specific cancer types (Table 2.1).

TABLE 2.1
List of Clinically Investigated Medicinal Plants with Their Mechanism of Action for Treatment of Cancer

S. No.	Biological Name	Common Name	Family	Mechanism of Action
1	*Zingiber officinale*	Ginger	Zingiberaceae	It is reported to be effective in ameliorating the side effects of γ-radiation and possess chemosensitization effects in certain neoplastic cells
2	*Curcuma longa* L.	Turmeric	Zingiberaceae	It inhibits invasion of tumour cells and *in vitro* metastasis by reducing MMP-2 activity and inhibiting cell invasion by hep2 (epidermoid carcinoma line)
3	*Phyllanthus amarus*	Jaramla	Phyllanthaceae	It decreases in *N*-nitrosodiethylamine (NDEA)-induced tumour incidence
4	*Munronia pinnata*	Bin kohomba	Meliaceae	It inhibits cancer cell lines
5	*Smilax zeylanica*	Kabarossa	Smilacacaea	Suppressed benzopyrene caused lung carcinoma by decreasing the number of nodules in the lung
6	*Tinospora cordifolia*	Moon seed	Menispermaceae	It is capable of inhibiting cell cycles
7	*Adenanthera pavonina*	Red breed tree, red sandalwood	Leguminosae	It exhibited anti-proliferative effects (cell survival inhibition)
8	*Thespesia populnea*	Tulip tree	Malvaceae	This plant has been found to inhibit neoplastic transformation in solid tumour
9	*Phyllanthus emblica*	Gooseberry	Phyllanthaceae	It regulates AP-1 inhibition and targets the transcription of viral oncogenes responsible for cervical cancer development and progression
10	*Catharanthus roseus*	Madagascar periwinkle,	Apocynaceae	They bind to tubulin and interrupt microtubular activity by halting cell cycle metaphase
11	*Taxus brevifolia*	Yew	Taxaceae	By stopping metaphase of the cell cycle, they bind to tubulin and interrupt microtubular activity
12	*Nothapodytes foetida*	Nothapodytes tree	Icacinaceae	This tree is used to treat leukaemia and anti-tumour disease
13	*Podophyllum peltatum*	Devil's apple, hog apple, Indian apple, umbrella plant, wild lemon	Berberidaceae	This agent is aimed at topoisomerase II and forms a complex with topoisomerase II and DNA induced by this complex break in double-stranded DNA
14	*Tabebuia impetiginosa, T. avellanedae*	Lapacho, Pau d'arco, Taheebo, and Ipe Roxo	Bignoniaceae	It is promoted for a number of human ailments, including cancer as a treatment

(*Continued*)

TABLE 2.1 (*Continued*)
List of Clinically Investigated Medicinal Plants with Their Mechanism of Action for Treatment of Cancer

S. No.	Biological Name	Common Name	Family	Mechanism of Action
15	*Cannabis sativa*	Marijuana, bhang, ganja, and hashish	Canabaceae	It is widely used as a psychoactive drug compound
16	*Camptotheca acuminata*	Xi shu, happy tree	Nyssaceae	This binds to topoisomerase and enables cleavage of DNA
17	*Betula alba*	Birch	Betulaceae	Efficient in handling patients suffering from breast cancer
18	*Colchicum autumnale*	Naked ladies, colchicum, and meadow saffron	Colchicaceae	It is used for the treatment of an inflammatory disorder. It is also value for its effects of chemotherapy
19	*Ziziphus nummularia*	Bhukamtaka sukhsharanphala, harbour	Rhamnaceae	Effectively kill cancer cells which are resistant to other chemical agents
20	*Andrographis paniculata*	Bhunimba and kalmegha, kiryat	Acanthaceae	It is a potent chemoprotective agent that acts against a variety of infectious and oncogenic agents
21	*Centella asiatica* L.	Brahmamanduki	Apiceae	It plays on cancer cells with the anti-proliferative effects
22	*Withania somnifera* L.	Ashwagandha	Solanaceae	It induces both intrinsic and extrinsic apoptosis signalling cascades in cancer cells, triggered by an increased production of reactive oxygen species (ROS) and nitric oxide (NO)
23	*Cedrus deodara*	Devdar	Pinaceae	Increase in apoptosis and post-apoptotic necrosis
24	*Boswellia serrata*	Torchwoods	Burseraceae	Efficient against brain tumours, it inhibits DNA synthesis and HL-60 cell development in different cancer cell lines
25	*Wedelia chinensis*	Sunflower	Asteraceae	It inhibits the signalling pathway for the androgen receptor (AR)
26	*Platycodon grandiflorum*	Bell flower	Campanulaceae	They were mainly used in lung cancer to reduce the toxicity and cancer-related symptoms associated with therapy and sometimes to directly increase anti-cancer effects
27	*Morus alba*	Mulberry	Moraceae	It induced arrest and apoptosis of cell growth in human cancer cells of colorectal origin
28	*Prunus armeniaca*	Apricot	Rosaceae	It shows inhibition of cancer cells by DNA damage and activation of apoptosis-inducing enzymes

(*Continued*)

TABLE 2.1 (*Continued*)
List of Clinically Investigated Medicinal Plants with Their Mechanism of Action for Treatment of Cancer

S. No.	Biological Name	Common Name	Family	Mechanism of Action
29	*Hemidesmus indicus*		Apocynaceae	It induces an immunogenic form of cell death in cancer cells of DLD1 (colon), stimulating upregulation of CD83 that triggers maturation of the dendritic cells
30	*Podophyllum peltatum*			These complex breaks in double-stranded DNA
31	*Boerhavia diffusa*	Hog weed	Nyctaginaceae	Anti-proliferative effects on HeLa cells with inhibition of cell cycle
32	*Tussilago farfara*	Coltsfoot		TF-induced cytotoxic and apoptotic activities by the floral section of *T. farfara* cells used in human colon cancer
33	*Capsicum annum* L.	Bell pepper or chilli pepper	Solanaceae	By inducing apoptosis, it is successful against gastric cancer, and modulates MAPK signalling
34	*Solanum melongena* L.	Brinjal	Solanaceae	It caused significant anti-proliferative effects against hepatic cancer cells by stopping the cell cycle at the S phase and inducing apoptosis
35	*Solanum nigrum* L.	Black night shade	Solanaceae	It has induced major anti-proliferative effects on hepatic cancer cells by blocking the S-phase cell cycle and inducing apoptosis
36	*Solanum tuberosum* L.	Potato	Solanaceae	It suppressed the growth of cancer cell proliferation
37	*Malus domestica*	Apples	Rosaceae	Apple peels showed potent anti-proliferative and anti-cancer activity
38	*Rosa canina* L.	Dog rose	Rosaceae	It showed inhibitory effect on human tumour cell proliferation by inducing apoptosis
39	*Rubus fruticosus*	Blackberry	Rosaceae	It inhibits the cell proliferation of cancer cell by inducing apoptosis in human leukaemia cell line
40	*Hibiscus sabdariffa* L.	Sorrel	Malvaceae	It showed effective anti-cancer activity on hepato carcinoma cell line
41	*Abelmoschus esculentus*	Ladies finger	Malvaceae	It inhibits the growth of cancer cells
42	*Mentha spicata*	Menthol	Lamiaceae	They exhibited great extent of cytotoxic effect (80%) against HeLa, Hep-2, and PC-3 cancer cell lines
43	*Plectranthus amboinicus*	Indian borage	Lamiaceae	It showed potent cytotoxic activity against HeLa cell line

(*Continued*)

TABLE 2.1 (*Continued*)
List of Clinically Investigated Medicinal Plants with Their Mechanism of Action for Treatment of Cancer

S. No.	Biological Name	Common Name	Family	Mechanism of Action
44	*Tectona grandis*	Teak	Lamiaceae	It showed potent anti–breast cancer activity against MCF-7 cells
45	*Ocimum sanctum*	Holy basil	Lamiaceae	It exhibits anti-cancer activity on fibro sarcoma cells. It also showed cytotoxic effect on cancer cells by decreasing cell multiplication, alternation in mitochondrial membrane potential, and increasing intracellular ROS and apoptosis in NC1-H460 cell line
46	*Acacia nilotica* L.	Babul	Fabaceaea	It exhibits potential anti-cancer activity through the growth inhibition, cell cycle arrest, and the apoptosis on MCF-7 and A549 cells
47	*Pisum sativum*	Green pea	Fabaceaea	It showed great anti-cancer activity against MCF-7 cells
48	*Rhus verniciflua*	Lacquer tree	Anacardiaceae	It increased the phosphorylation of AMP-activated protein kinase (AMPK) and downstream acetyl-CoA carboxylase (ACC) and inhibited the AMPK-dependent mode of cell viability
49	*Perilla frutescens*	Perilla, Beafsteak plant	Labiatae	It induces apoptosis and G1 phase arrest of HL-60 cells in human leukaemia through combinations of medicated, mitochondrial, and endoplasmic reticulum stress-induced pathways
50	*Stemona japonica*	Stemona root	Stemonaceae	Topoisomerase II inhibitors in DNA are generally potent, and some also induce apoptosis in cancer cells
51	*Draba nemorosa*	Wood whitelow grass	Brassicaceae	It normally exerts anti-cancer activity by cell cycle arrest and apoptosis induction
52	*Scutellaria baicalensis*	Baikal skull cap	Lamiaceae	Inhibit enzymatic synthesis of eicosanoids
53	*Allium cepa*	Onion basal	Amaryllidaceae	Polyphenols isolated from *Allium cepa* L. It induces apoptosis by induction of p53 and suppression of Bcl-2 through inhibiting PI3K–Akt signalling pathway in AGS human cancer cells

(*Continued*)

TABLE 2.1 (*Continued*)
List of Clinically Investigated Medicinal Plants with Their Mechanism of Action for Treatment of Cancer

S. No.	Biological Name	Common Name	Family	Mechanism of Action
54	*Allium sativum* L.	Garlic/ thorn	Amaryllidaceae	It has significant anti-tumoural and antimotility effects on MDA-MB-231 and MCF-7 human breast cancer cells which are attributed to its bioactive molecules
55	*Mangifera indica* L.	Mango	Anacardiaceae	The efficacy of *M. indica* extracts from all mango varieties in the inhibition of cell growth was tested in SW-480 colon carcinoma cells
56	*Pistacia palaestina boiss*	Mastic tree, lentisk	Anacardiaceae	*P. palaestina* significantly reduced CXCL8 levels, a chemokine contributing in CRC cells proliferation and metastasis
57	*Annona muricata* L.	Soursop/ keshta	Annonaceae	*A. muricata* extract was found to suppress the proliferation of HL-60 cells by inducing morphology changes, G0/G1 phase cell detention, damage to cell viability, and detriment of membrane mitochondrial potential
58	*Daucus guttatus* Sm.	Wild carrot, jazar barry	Apiaceae	Daucus carota pentane-based fractions arrest the cell cycle and increase apoptosis in MDA-MB-231 breast cancer cells
59	*Petroselinum crispum*	Fuss/ parsley/ bokdonas	Apiaceae	AP-02-induced anti-tumour effect on COLO 205 xenograft tumour growth formation. Researchers demonstrate that AP-02 inhibits cancer cell proliferation through G0/G1 cell cycle arrest
60	*Calotropis procera*	Dryand./ apple of Sodom, (Mudar)/ a'oshar Basek	Apocynaceae	A significant cytotoxicity and morphological changes were observed post-treatment with *C. procera* extract in MCF-7 breast cancer cells
61	*Nerium oleander* L.	Oleander/ dafla	Apocynaceae	It acts against MCF-7 breast cancer cell line which revealed that the stabilized AuNPs were highly effective for the apoptosis of cancer cells
62	*Arum dioscoridis* Sm.	Spotted arum/ loof mobarkash	Araceae	It arrests the cell cycle and increase apoptosis in MDA-MB-231 breast cancer cells

(*Continued*)

TABLE 2.1 (*Continued*)
List of Clinically Investigated Medicinal Plants with Their Mechanism of Action for Treatment of Cancer

S. No.	Biological Name	Common Name	Family	Mechanism of Action
	Polygonatum multiflorum L.	David's harp/ khatem soleyman	Asparagaceae	It inhibits human breast cancer cells (MCF-27) and Herps Eac tumour masses and human oesophageal cancer ECA-109 cells, human gastric cancer HGC-27 cells
63	*Brassica oleracea* L.	Cabbage/ malfof	Brassicaceae	
64	*Sinapis arvensis*	Mustard/ khardal	Brassicaceae	Bone
65	*Capparis spinosa* L.	Caperbush/ coba	Capparaceae	Bone cancer
66	*Colchicum hierosolymitanum* L.	Colchicum/ lohlah	Colchicaceae	Skin
67	*Achillea aleppica* DC.	Yarrow/ kaysoom	Compositae	Liver
68	*Cichorium endivia* L.	Chicory/ shoka	Compositae	Stomach and colon
69	*Inula viscosa* L.	Aiton/ false yellowhead/ tayon	Compositae	Kidney and bladder
70	*Onopordum cynarocephalum*	Artichoke cotton-thistle/ kondrees	Compositae	Extracts of lignans and sesquiterpenes are known to be inducers of cytotoxicity by triggering apoptosis
71	*Silybum marianum* L.	Gaertn./ milk thistle khorfeesh	Asteraceae	Silymarin's anti-cancer effects by causing cell cycle arrest and inducing apoptosis in different types of cancer
72	*Taraxacum syriacum boiss*	Dandelion/ hindeba	Compositae	Suppress and arrest the G0/G1-phase, reduces DNA synthesis, and induces apoptosis
73	*Citrullus colocynthis* L.	Chrad./ bitter gourd/ hantha	Cucurbitaceae	It has potential chemotherapeutic agent against MCF-7 and AGS cell lines in gastric adenocarcinoma and breast cancer, respectively
74	*Cucumis sativus* L.	Cucumber/ kheyar	Cucurbitaceae	Exposure of Cucurbitin-1 to nasopharyngeal carcinoma cells *in vitro* clonogenicity and *in vivo* tumourigenicity caused decrease in nasopharyngeal carcinoma cell
75	*Ecballium elaterium* L.	Cucumber/ ketha' alhemar	Cucurbitaceae	It suppresses the proliferation of human breast adenocarcinoma cells (MDA-MB468)
76	*Ephedra alata decne*	Ephedra alata decne	Ephedraceae	It showed anti-proliferative and pro-apoptotic action against the MCF-7 human breast cancer cell line
77	*Arbutus andrachne* L.	Strawberry tree/ kotlob	Ericaceae	Stomach
78	*Euphorbia hierosolymitana boiss*	Spurge/ halablabon	Euphorbiaceae	Induces cancer cell death by apoptosis and inhibits proliferation of cancer cell

(*Continued*)

TABLE 2.1 (*Continued*)
List of Clinically Investigated Medicinal Plants with Their Mechanism of Action for Treatment of Cancer

S. No.	Biological Name	Common Name	Family	Mechanism of Action
79	*Quercusc Calliprinos webb*	Palestine oak/ baloo	Fagaceae	Colorectal
80	*Quercus ithaburensis decne*	Valonia oak/ sendnyan	Fagaceae	Skin
81	*Hypericum perforatum* L.	Wort/ oshbat ala'ran	Hypericaceae	Brain
82	*Crocus sativus* L.	Saffron/ za'faran	Iridaceae	Liver and kidney
83	*Melissa officinalis* L.	Lemon Balm mint/ torenjan	Lamiaceae	Lung and non-Hodgkin lymphoma
84	*Origanum jordanicum*	Danin & kunne/ Thyme/ za'ata	Lamiaceae	Lung, throat cancer
85	*Rosmarinus officinalis* L.	Rosemary/ hasa alban	Lamiaceae	Arrests multiplication of Cancerous cells by breaking down the microtubule
86	*Salvia fruticosa Mil*	Maryamya	Lamiaceae	Colon and liver
87	*Salvia palaestina*	Benth./ kosaen (kharna)	Lamiaceae	Brain
88	*Teucrium capitatum* L.	Teucrium/ jada	Lamiaceae	Pancreatic and liver
89	*Lycium europaeum*	Box thorn	Solanaceae	Bladder, prostate, and breast
90	*Verbascum sinuatum* L.	Mullein	Scrophulariaceae	Breast
91	*Acer obtusifolium* Sm.	Syrian	Sapindaceae	Throat and lung
92	*Salix alba* L.	White Salix/ Sofsaf abyad	Salicaceae	Colon
93	*Crataegus azarolus* L.	Azarole hawthorn	Rosaceae	Lung
94	*Nigella arvensis* L.	Black cumin	Ranunculaceae	Lung, brain, and skin
95	*Cyclamen persicum Mill*	Cyclamen/ Sapoon alraa'e	Primulaceae	Prostate and bladder
96	*Triticum aestivum* L.	Bread wheat	Poaceae	Colon
97	*Plantago lanceolata* L.	Narrow leaf plantain/ Lesan alhama	Plantaginaceae	Throat
98	*Abelmoschus esculenthus*	Bhindi	Malvaceae	It suppressed cancer cell growth and helps in preventing cancer and acts against three carcinoma cell lines (MCF-7, HeLa, and HepG2)
99	*Arum palaestinum Boiss*	Cuckoo pint/ loof	Araceae	It acts against four human carcinoma cell lines, Hep2, HeLa, HepG2, and MCF7, and helps to identify the volatile components which may be responsible for the potential anti-tumour activity

LIST OF ABBREVIATIONS

24-epiBL	24-epibrassinolide
28-homoCS	28-homocastasterone
AIF	Alpinumi
APC	ATP-binding cassette
BRs	Brassinosteroids
CDKs	Cyclin-dependent kinases
CSF-1	Macrophage colony-stimulating factor-1
EGFR	Epidermal growth factor receptor
FGF-1 and FGF-2	Fibroblast growth factors
GAP	GTPase-activating protein
HER2	Human epidermal growth factor receptor 2
HGF	Hepatocyte growth factor
HL	Human leukaemia
IGF-1	Insulino-like growth factor-1
MAPK	Mitogen-activated protein kinase
MLF	4′-Methoxy licoflavanone
NAG-1	NSAID activated gene
NF-κB	Nuclear factor kappa B
NGF	Nerve growth factor
OR	Oestrogen receptor
PDGF	Platelet-derived growth factor
PDT	Photodynamic treatment
PTKs	Protein tyrosine kinases
ROS	Reactive oxygen species
SM	Secondary metabolite
STAT proteins	Signal transducer and activator of transcription
TM	Traditional medicine
TNF	Tumour necrosis factor

BIBLIOGRAPHY

Adhikari, S., Dongol, R., Hewett, Y., Shah, B. K., Joseph, S., & Medical, R. (2014). Vincristine-induced blindness: A case report and review of literature. *Anticancer Research, 6734*, 6731–6733.

Afshari, J. T., Brook, A., & Mousavi, S. H. (2008). Study of cytotoxic and apoptogenic properties of saffron extract in human cancer cell lines. *Food and Chemical Toxicology, 46*, 3443–3447.

Aggarwal, S., Takada, Y., Singh, S., Myers, J. N., & Aggarwal, B. B. (2004). Inhibition of growth and survival of human head and neck squamous cell carcinoma cells by curcumin via modulation of nuclear factor-κB signalling. *International Journal of Cancer, 111*, 679–692.

Amantini, C., Mosca, M., Nabissi, M., Lucciarini, R., Caprodossi, S., Arcella, A., Giangaspero, F., & Santoni, G. (2007). Capsaicin-induced apoptosis of glioma cells is mediated by TRPV1 vanilloid receptor and requires p38 MAPK activation. *Journal of Neurochemistry, 102*, 977–990.

Amruthraj, N. J., Preetam Raj, J. P., Saravanan, S., & Lebel, L. A. (2014). *In vitro* studies on anticancer activity of capsaicinoids from capsicum Chinese against human hepatocellular carcinoma cells. *International Journal of Pharmacy and Pharmaceutical Sciences, 6*, 254–558.

Anand, P., Sundaram, C., Jhurani, S., Kunnumakkara, A. B., & Aggarwal, B. B. (2008). Curcumin and cancer: An "old-age" disease with an "age-old" solution. *Cancer Letters, 267*, 133–164.

Araujo, C. C., & Leon, L. L. (2001). Biological activities of *Curcuma longa* L. *Memórias do Instituto Oswaldo Cruz, 96*, 723–728.

Arnold, J. T., Wilkinson, B. P., Sharma, S., & Steele, V. E. (1993) Evaluation of chemopreventative agents in different mechanistic classes using a rat epithelial cell culture transformation assay. *Cancer Research, 73*, 537–543.

Aruna, S. J., Benjamin, P. M., Nirupama, G., & Satya, N. (2002). β-Catenin-mediated transactivation and cell–cell adhesion pathways are important in curcumin (diferuloylmethane)-induced growth arrest and apoptosis in colon cancer cells. *Oncogene, 21*, 8414–8427.

Aung, H. H., Wang, C. Z., Ni, M., Fishbein, A., Mehendale, S. R., & Xie, J. T. (2007). Crocin from *Crocus sativus* possesses significant anti-proliferation effects on human colorectal cancer cells. *Experimental Oncology, 29*, 175–180.

Bakshi, H., Sam, S., & Rozati, R. (2010) DNA fragmentation and cell cycle arrest: A hallmark of apoptosis induced by crocin from Kashmiri saffron in a human pancreatic cancer cell line. *Asian Pacific Journal of Cancer Prevention, 11*, 675–679.

Bharti, B., & Alok, B. A. (2002). Nuclear factor-kappa B and cancer: Its role in prevention and therapy. *Biochemistry & Pharmacology, 64*, 883–888.

Blank, F., & Muggia, S. (2013). Drug evaluation a pharmacokinetic evaluation of topotecan as a cervical cancer therapy. *Expert Opinion on Drug Metabolism & Toxicology, 9*, 215–224.

Chabner, B., & Roberts, T. G. (2005). Timeline: Chemotherapy and the war on cancer. *Nature Reviews: Cancer, 5*, 65–72.

Cooper, G. M. (2000). *The cell: A molecular approach* (2nd ed.). Sinauer Associates.

Debatin, K. M. (2004). Apoptosis pathways in cancer and cancer therapy. *Cancer Immunology, Immunotherapy, 53*, 153–159.

DeBono, A., Capuano, B., & Scammells, P. J. (2015). Progress toward the development of noscapine and derivatives as anticancer agents. *Journal of Medicinal Chemistry, 58*, 5699–5727.

Delgoda, R., & Murray, J. E. (2017). Evolutionary perspectives on the role of plant secondary metabolites. In S. Badal & R. Delgoda (Eds.), *Pharmacognosy: Fundamentals, applications and strategies* (1st ed., pp. 93–100). Academic Press.

Devriese, L. A., Witteveen, P. E., Mergui-Roelvink, M., Smith, D. A., Lewis, L. D., Mendelson, D. S., Bang, Y. J., Chung, H. C., Dar, M. M., Huitema, A. D., Beijnen, J. H., Voest, E. E., & Schellens, J. H. M.(2014). Pharmacodynamics and pharmacokinetics of oral topotecan in patients with advanced solid tumours and impaired renal function. *British Journal of Clinical Pharmacology, 80*, 253–266.

Dickson, M. A., & Schwartz, G. K. (2009). Development of cell-cycle inhibitors for cancer therapy. *Current Oncology, 16*, 36–43.

Elmore, S. (2007). Apoptosis: A review of programmed cell death. *Toxicologic Pathology, 35*, 459–516.

Evans, A. E., Farber, S., Brunet, S., & Mariano, P. J. (1963). Vincristine in the treatment of acute leukaemia in children. *Cancer, 16*, 1302–1306.

Fan, Y., Patima, A., Chen, Y., Zeng, F., He, W., Luo, L., Jie, Y., Zhu, Y., Zhang, L., Lei, J., Xie, X., & Zhang, H. (2015). Cytotoxic effects of beta-carboline alkaloids on human gastric cancer SGC-7901 cells. *International Journal of Clinical and Experimental Medicine, 8*, 12977–12982.

Foster, D. A., Yellen, P., Xu, L., & Saqcena, M. (2010). Regulation of G1 cell cycle progression: Distinguishing the restriction point from a nutrient-sensing cell growth checkpoint(s). *Genes Cancer, 1*, 1124–1131.

Fruman, D. A., & Rommel, C. (2014). PI3K and cancer: Lessons, challenges and opportunities. *Nature Reviews. Drug Discovery, 13*, 140–156.

Ghavami, S., Hashemi, M., Ande, S. R., Yeganeh, B., Xiao, W., Eshraghi, M., Bus, C. J., Kadkhoda, K., Wiechec, E., Halayko, A. J., & Los, M.(2009). Apoptosis and cancer: Mutations within caspase genes. *Journal of Medical Genetics, 46*, 497–510.

Giacinti, C., & Giordano, A. (2006). RB and cell cycle progression. *Oncogene, 25*, 5220–5227.

Gourmelon, C., Bourien, H., Augereau, P., Patsouris, A., Frenel, J.-S., & Campone, M. (2016). Vinflunine for the treatment of breast cancer. *Expert Opinion on Pharmacotherapy, 17*, 1817–1823.

Greenbaum, L. E. (2016). Cell cycle regulation and hepatocarcinogenesis. *Cancer Biology & Therapy, 3*, 1200–1207.

Guo, Z. (2017). The modification of natural products for medical use. *Acta Pharmaceutica Sinica. B, 7*, 119–136.

Habli, Z., Toumieh, G., Fatfat, M., Rahal, O. N., & Gali-Muhtasib, H. (2017). Emerging cytotoxic alkaloids in the battle against cancer: Overview of molecular mechanisms. *Molecules, 22*, 250.

Hartley, J. A., Hochhauser, D., Boone, J. J., Bhosle, J., & Tilby, M. J. (2009). Involvement of the HER2 pathway in repair of DNA damage produced by chemotherapeutic agents. *Molecular Cancer Therapeutics, 8*, 3015–3023.

Holmsten, K., Dohn, L., Jensen, N. V., Shah, C. H., Jäderling, F., Pappot, H., & Ullén, A. (2016). Vinflunine treatment in patients with metastatic urothelial cancer: A Nordic retrospective multicenter analysis. *Oncology Letters, 12*, 1293–1300.

Jackson, S. P., & Bartek, J. (2010). The DNA-damage response in human biology and disease. *Nature, 461*, 1071–1078.

Jeong, C. H., & Joo, S. H. (2016). Downregulation of reactive oxygen species in apoptosis. *Journal of Cancer Prevention, 21*, 13–20.

Kabera, J. N., Semana, E., Mussa, A. R., & He, X. (2014). Plant secondary metabolites: Biosynthesis, classification, function and pharmacological properties. *Journal of Pharmaceutics & Pharmacology, 2*, 377–392.

Safia, Kamil, M., Jadiya, P., Sheikh, S., Haque, E., Nazir, A., Lakshmi, V., & Mir, S. S. (2015). The chromone alkaloid, rohitukine, affords anti-cancer activity via modulating apoptosis pathways in A549 cell line and yeast mitogen activated protein kinase (MAPK) pathway. *PLoS One, 10*, 1–18.

Kampan, N. C., Madondo, M. T., McNally, O. M., Quinn, M., & Plebanski, M. (2015). Paclitaxel and its evolving role in the management of ovarian cancer. *BioMed Research International, 2015*, 413076.

Laryea, D., Isaksson, A., Wright, C. W., Larsson, R., & Nygren, P. (2009). Characterization of the cytotoxic activity of the indoloquinoline alkaloid cryptolepine in human tumour cell lines and primary cultures of tumour cells from patients. *Investigational New Drugs, 27*, 402–411.

Lee, S. T., Wong, P. F., & Cheah, S. C. (2011). Alpha-tomatine induces apoptosis and inhibits nuclear factor-kappa B activation on human prostatic adenocarcinoma PC-3 cells. *PLoS One, 6*, e18915.

Li, M., Li, P., Zhang, M., Ma, F., & Su, L. (2014). Brucine inhibits the proliferation of human lung cancer cell line PC-9 via arresting cell cycle. *Zhongguo Fei Ai Za Zhi, 17*, 444–450.

Li, L., Xu, Y., & Wang, B. (2014). Liriodenine induces the apoptosis of human laryngocarcinoma cells via the upregulation of p53 expression. *Oncology Letters, 15*, 1121–1127.

Liew, S. Y., Looi, C. Y., Paydar, M., Cheah, F. K., Leong, K. H., Wong, W. F., Mustafa, M. R., Litaudon, M., & Awang, K. (2014). Subditine, a new monoterpenoid indole alkaloid from bark of *Nauclea subdita* (Korth.) Steud. induces apoptosis in human prostate cancer cells. *PLoS One*, *9*, e87286.

Lou, C., Takahashi, K., Irimura, T., Saiki, I., & Hayakawa, Y. (2014). Identification of hirsutine as an anti-metastatic phytochemical by targeting NF-κB activation. *International Journal of Oncology*, *45*, 2085–2091.

Lou, C., Yokoyama, S., Saiki, I., & Hayakawa, Y. (2015). Selective anticancer activity of hirsutine against HER2 positive breast cancer cells by inducing DNA damage. *Oncology Reports*, *33*, 2072–2076.

Ma, B., & Hottiger, M. O. (2016). Crosstalk between Wnt/β-catenin and NF-κB signaling pathway during inflammation. *Frontiers in Immunology*, *7*, 378.

Mansoor, T. A., Borralho, P. M., Dewanjee, S., Mulhovo, S., Rodrigues, C. M. P., & Ferreira, M. J. (2013). Monoterpene bisindole alkaloids, from the African medicinal plant *Tabernaemontana elegans*, induce apoptosis in HCT116 human colon carcinoma cells. *Journal of Ethnopharmacology*, *149*, 463–470.

Matsui, T. A., Sowa, Y., Murata, H., Takagi, K., Nakanishi, R., Aoki, S., Yoshikawa, M., Kobayashi, M., Sakabe, T., & Kubo, T. (2007). The plant alkaloid cryptolepine induces p21WAF1/CIP1 and cell cycle arrest in a human osteosarcoma cell line. *International Journal of Oncology*, *31*, 915–922.

Mcilwain, D. R., Berger, T., & Mak, T. W. (2013). Caspase functions in cell death and disease. *Cold Spring Harbor Perspectives in Biology*, 5, a008656.

Neidle, S. (2016). Quadruplex nucleic acids as novel therapeutic targets. *Journal of Medicinal Chemistry*, *59*, 5987–6011.

Nicum, S. J., & O'Brien, M. E. (2007). Topotecan for the treatment of small-cell lung cancer. *Expert Review of Anticancer Therapy*, *7*, 795–801.

Nordin, N., Majid, N. A., Hashim, N. M., Rahman, M. A., Hassan, Z., & Ali, H. M. (2015). Liriodenine, an aporphine alkaloid from *Enicosanthellum pulchrum*, inhibits proliferation of human ovarian cancer cells through induction of apoptosis via the mitochondrial signaling pathway and blocking cell cycle progression. *Drug Design, Development and Therapy*, *9*, 1437–1448.

Nwodo, J. N., Ibezim, A., Simoben, C. V., & Ntie-Kang, F. (2016). Exploring cancer therapeutics with natural products from African medicinal plants, Part II: Alkaloids, terpenoids and flavonoids. *Anticancer Agents Medicinal Chemistry*, *16*, 108–127.

O'Brien, M. E. R., Eckardt, J., & Ramlau, R. (2007). Recent advances with topotecan in the treatment of lung cancer. *Oncologist*, *12*, 1194–1204.

Okamoto, D. Y., & Okamoto, K. (2010). Structural insights into G-quadruplexes: Towards new anticancer drugs. *Future Medicinal Chemistry*, *2*, 619–646.

Prokhorova, E. A., Zamaraev, A. V., Kopeina, G. S., Zhivotovsky, B., & Lavrik, I. N. (2015). Role of the nucleus in apoptosis: Signaling and execution. *Cellular and Molecular Life Sciences*, *72*, 4593–4612.

Sajadian, S., Vatankhah, M., Majdzadeh, M., Kouhsari, S. M., Ghahremani, M. H., & Ostad, S. N. (2015). Cell cycle arrest and apoptogenic properties of opium alkaloids noscapine and papaverine on breast cancer stem cells. *Toxicology Mechanisms and Methods*, *25*, 388–395.

Saraste, A., & Pulkki, K. (2000). Morphologic and biochemical hallmarks of apoptosis. *Cardiovascular Research*, *45*, 528–537.

Schelz, Z., Ocsovszki, I., Bózsity, N., Hohmann, J., & Zupko, I. (2016). Antiproliferative effects of various furanoacridones isolated from *Ruta graveolens* on human breast cancer cell lines. *Anticancer Research*, *36*, 2751–2758.

Shih, Y. W., Shieh, J. M., Wu, P. F., Lee, Y. C., & Chen, Y. Z. (2009). Alpha-tomatine inactivates PI3K/Akt and ERK signaling pathways in human lung adenocarcinoma A549 cells: Effect on metastasis. *Food and Chemical Toxicology*, *47*, 1985–1995.

Shu, G., Mi, X., Cai, J., Zhang, X., Yin, W., Yang, X., Li, Y., Chen, L., & Deng, X. (2013). Brucine, an alkaloid from seeds of *Strychnos nux-vomica* Linn., represses hepatocellular carcinoma cell migration and metastasis: The role of hypoxia inducible factor 1 pathway. *Toxicology Letters*, *222*, 91–101.

Sundaram, V. (2006). RTK/Ras/MAPK signaling. *WormBook*, 1–19.

Uche, F. I., Drijfhout, F. P., McCullagh, J., Richardson, A., & Wen, L. W. (2016). Cytotoxicity effects and apoptosis induction by bisbenzylisoquinoline alkaloids from *Triclisia subcordata*. *Phytotherapy Research*, *30*, 1533–1539.

Visconti, R., Della Monica, R., & Grieco, D. (2016). Cell cycle checkpoint in cancer: A therapeutically targetable double-edged sword. *Journal of Experimental & Clinical Cancer Research*, *35*, 153.

Wang, K. B., Li, D. H., Hu, P., Wang, W. J., Lin, C., Wang, J., Lin, B., Bai, J., Pei, Y. H., Jing, Y.-K., Li, Z.-L., Yang, D., & Hua, H.-M. (2016). A series of β-carboline alkaloids from the seeds of *Peganum harmala* show G-quadruplex interactions. *Organic Letters*, 18, 3398–3401.

Wang, X. D., Li, C. Y., Jiang, M. M., Li, D., Wen, P., Song, X., Chen, J. D., Guo, L. X., Hu, X. P., Li, G. Q., Zhang, J., Wang, C.-H., & He, Z.-D. (2016). Induction of apoptosis in human leukemia cells through an intrinsic pathway by cathachunine, a unique alkaloid isolated from *Catharanthus roseus*. *Phytomedicine*, 23, 641–653.

Waziri, P. M., Abdullah, R., Yeap, S. K., Omar, A. R., Kassim, N. K., Malami, I., How, C. W., Etti, I. C., & Abu, M. L. (2016). Clausenidin induces caspase-dependent apoptosis in colon cancer. *BMC Complementary and Alternative Medicine*, *16*, 256.

Weaver, B. A. (2014). How Taxol/paclitaxel kills cancer cells. *Molecular Biology of the Cell*, *25*, 2677–2681.

Weinberg, R. A. (2013). *The biology of cancer* (2nd ed.). Garland Science.

Wink, M., Ashour, M. L., & El-Readi, M. Z. (2012). Secondary metabolites inhibiting ABC transporters and reversing resistance of cancer cells and fungi to cytotoxic and antimicrobial agents. *Frontiers in Microbiology*, *3*, 1–15.

Yang, X. K., Xu, M. Y., Xu, G. S., Zhang, Y. L., & Xu, Z. (2014). *In vitro* and *in vivo* antitumor activity of scutebarbatine A on human lung carcinoma A549 cell lines. *Molecules*, *19*, 8740–8751.

Yao, H., Liu, J., Xu, S., Zhu, Z., & Xu, J. (2017). The structural modification of natural products for novel drug discovery. *Expert Opinion on Drug Discovery*, *12*, 121–140.

Zeina, H., Georgio, T., Maamoun, F., & Omar, N. R. (2017). Emerging cytotoxic alkaloids in the battle against cancer: Overview of molecular mechanisms. *Molecules*, *22*, 25.

Zhang, J. F., Liu, J., Wang, Y., & Zhang, B. (2016). Novel therapeutic strategies for patients with triple-negative breast cancer. *Onco Targets and Therapy*, *9*, 6519–6528.

Zheng, L., Wang, X., Luo, W., Zhan, Y., & Zhang, Y. (2013). Brucine, an effective natural compound derived from nux-vomica, induces G1 phase arrest and apoptosis in LoVo cells. *Food and Chemical Toxicology*, *58*, 332–339.

Zhu, H., & Gooderham, N. J. (2006). Mechanisms of induction of cell cycle arrest and cell death by cryptolepine in human lung adenocarcinoma A549 cells. *Toxicological Sciences*, *91*, 132–139.

Zupkó, I., Réthy, B., Hohmann, J., Molnár, J., Ocsovszki, I., & Falkay, G. (2009). Antitumor activity of alkaloids derived from Amaryllidaceae species. *In Vivo (Athens, Greece)*, *23*, 41–48.

3 Medicinal Plants Having Anti-cancer Activity

Sapna Malviya, Neelesh Malviya, Varsha Johariya, Rajiv Saxena, Ruchi Gupta, Manisha Dhere, Anindya Goswami, Ankur Joshi, and Anamika Singh

CONTENTS

DOI: 10.1201/9781003251712-3

3.1 *CATHARANTHUS ROSEUS*

3.1.1 BACKGROUND

C. roseus (L.) is also referred to as Madagascar periwinkle. The plant is potentially recognized as possessing medicinal properties. The presence of two important anti-tumour terpenoids, including indole alkaloids, vincristine, and vinblastine, makes the plant medicinally important. *C. roseus* (L.) is an evergreen plant and ornamentally cultivated and has a rich aesthetic value.

C. roseus Linn. is best cultivated as an annual bedding plant in the presence of well-drained sandy loams. It needs full sun to part shade along with regular moisture. Overhead watering is commonly avoided. The cuttings are taken from plants at the time of late summer for overwintering in order to provide stock at the time of spring. The botanical classification of *C. roseus* is described below:

Kingdom	**Plantae**
Phylum	Magnoliophyta
Class	Magnoliopsida
Order	Gentianales
Family	Apocynaceae
Genus	*Catharanthus*
Species	*roseus*
Binomial name	*Catharanthus roseus* L.

3.1.2 ETHNOMEDICAL CONSIDERATIONS

The plant belongs to the *roseus* species of the genus *Catharanthus* in the family Apocynaceae. The plant is available throughout the world and found to possess alkaloids, flavonoids, and steroids as active chemical constituents. *C. roseus* Linn. is an inhabitant geographically of the Indian Ocean Island of Madagascar (Figure 3.1). The plant is now common in most of the tropical and subtropical regions throughout the world.

3.1.3 MORPHOLOGY AND PHARMACOGNOSTIC CHARACTERISTICS

3.1.3.1 Macroscopic Characteristics

The leaves of *C. roseus* are arranged opposite and are simple, petiolate, glabrous, or softly pubescent. The lamina is elliptic and found to be obovate or oblong-elliptic. The base of the leaves is cuneate or subcuneate and is often found oblique and slightly decurrent. The margin is entire and may be hairy and found to be membranous or thinly conspicuous. The upper surface of the leaves is puberulous or glabrescent and has dark shining green colour. The lower surface of the leaves is found to be pubescent or glabrous. The colour is light green and lateral nerves are rather close and arcuate.

FIGURE 3.1 Photographs of *Catharanthus roseus* Linn. (A) Flower bud, (B) Flower (C) *Catharanthus roseus* Linn plant, (D) Fruit and seeds of *Catharanthus roseus* Linn, and (E) Dried powder of flowers.

3.1.3.2 Microscopic Characteristics

The leaves of *C. roseus* are dorsiventral. The sinuous-walled lower epidermal cells are seen and the upper epidermal cells are slightly sinuous-walled or maybe walls slightly curved. The stomata available on the lower surface are mainly of the ranunculaceous type. The transverse section (TS) of the petiole has epidermal cells and is found to have thin walls along with a thick cuticle available in the outer walls. The cells of *C. roseus* have different sizes and shapes; they are either oval or may be elliptical in shape. The hypodermis consists of thin-walled parenchymatous cells followed by the cortex. The upper epidermal cells are much-elongated palisade and seen to be filled with chloroplasts. The lower epidermal cells have a similar size and shape as compared to the upper epidermis. In between the palisade cells and the lower epidermal cells, the spongy arenchymatous cells are observed. The cells are found to be oval, elliptical, oblong, or polygonal in shape.

3.1.4 Phytochemistry

C. roseus Linn. predominantly showed the presence of two pharmacologically important as well as active cytotoxic dimeric alkaloids—vinblastine and vincristine. Both are used for cancer chemotherapy. These cytotoxic dimeric alkaloids are present in extremely low yields, especially in leaves. Apart from this, the plant is shown to have carbohydrates, flavonoids, saponins, and alkaloids. The studies suggested that more than around 130 indole alkaloids are present in plants and the highest number can be identified at the flowering stage. The chief phytoconstituent present in the aerial part is alkaloids, including the following:

- Vincaleukoblastine
- Leurocristine, vincaleurocristine
- Vincarodine
- Leurocolombine
- Viramidine
- Vincathicine
- Isositsirikine
- Lochrovicine
- Catharanthine
- Vindolinine etc.

The main alkaloids present in the roots of *C. roseus* Linn. include Ajmalicine (raubasine), serpentine, and reserpine. Its flowers predominantly have coronaridine, tetrahydroalstonine, ajmalicine, vindorosidine, and vincristine, and its basal stems are shown to possess ajmalicine, vinceine, vineamine, raubasine, reserpine, catharanthine, etc. The flower also contains rosinidin, an anthocyanin pigment.

3.1.5 Pharmacological Properties

Catharanthus has significant anti-cancer activity against numerous cell types. Many parts of the plant are used in traditional systems of medicine for the effective treatment of many diseases. It is employed for the treatment of antidiabetic agents, anti-galactagogue, malaria, and skin infections. The plant has wide applications as its extracts have been used for diseases and disorders like ocular inflammation, diabetes, haemorrhage, and cancers. Its greatest activity is seen against multi-drug-resistant tumour types.

3.1.6 Scientific Investigation for the Management of Cancer

Arora (1998) and Arora et al. (2005) have studied and explored biotechnological routes for enhanced production of medicinally useful *Catharanthus* alkaloids. Wargin and Lucas (1994) have studied the alkaloids of *Catharanthus* and shown that these are integral parts of the chemotherapy regimen that can be used singly or in a combination of different alkaloids. Jordan et al. (1991) and Wilson et al. (1999) studied the anti-tumour effect of vinorelbine and vinflunine by binding to tubulin. Depierre et al. (1991),

Bennouna et al. (2006), and Mano (2006) observed that both vinorelbine and vinflunine are found to be effective in treating non-small cell lung cancer, breast cancer, and bladder cancer. Simoens et al. (2008) studied radiation-resistant tumours sensitive to radiation therapy. More studies and future investigations were suggested by the researchers.

3.1.7 Reported Mechanism of Action as Anti-cancer Activity

Vinblastine attaches to tubulin and thereby prevents the formation of microtubules. It is a crucial part of numerous chemotherapy regimens, including ABVD (adriamycin, bleomycin, vinblastine, and dacarbazine) for Hodgkin's lymphoma. It is cell cycle–specific to the M phase. The preferred treatment plan for metastatic testicular cancer includes vinblastine. It has been demonstrated that vinblastine fights cancer by obstructing glutamic acid metabolism. Additionally, Kaposi's sarcoma, mycosis fungoides, and breast cancer are all treated with vinblastine.

Vincristine inhibits the formation of microtubule structures by attaching to the structural protein tubulin dimer. The cell cycle's metaphase stage is where mitosis is stopped when the microtubules are disrupted. All sorts of rapidly dividing cells, including cancer cells, intestinal epithelium, and bone marrow, are affected by the *Catharanthus* alkaloids. *Vinca* alkaloids are also referred to as mitotic spindle poisons. The reason is it inhibits the assembly of the spindle forms from microtubules.

Vinca alkaloids inhibit mitosis in the cell cycle. Vincristine and vinblastine are scientifically explored to inhibit cancer cell growth at the time of metaphase that leads to the death of cells. The enhancement of apoptosis by *Vinca* alkaloids occurs with increasing concentrations of the cellular tumour antigen p53 along with cyclin-dependent kinase inhibitors. The increased levels of p53 and p21 cause the change in protein kinase activity.

3.1.8 Toxicological Aspects

The side effects of vinblastine include gastrointestinal toxicity, significant vesicant (blister-forming) activity, extravasation damage, and bone marrow suppression (which forms deep ulcers). This medication is not recommended for those with bacterial illnesses. Vinblastine should not be taken during pregnancy as it has been demonstrated to be embryotoxic, mutagenic, and carcinogenic in research on animals. Because it might be secreted into breast milk, breastfeeding is also not advised. Constipation, hyponatremia, peripheral neuropathy, and hair loss are the primary adverse effects of vincristine. Intrathecal ingestion of *Catharanthus* alkaloids has the potential to be fatal. There are cases of ascending paralysis brought on by severe encephalopathy and demyelination of the spinal nerves, which is accompanied by unbearable suffering and ultimately results in death.

3.1.9 Conclusion and Future Aspects

C. roseus is found all over the world and its alkaloids have been used for centuries in traditional medicine to treat a wide range of illnesses. *C. roseus* produce several indole alkaloids named *Vinca* alkaloids, which are widely used as antimitotic drugs in the treatment of cancer. *C. roseus* has the ability to cultivate in submerged culture using bioreactors. *C. roseus* is considered one of the plant biofactors for the production of highly potent anti-cancer molecules. Therefore, cultivation and production of *C. roseus* are easy for the production of anti-cancer compounds in high concentration, in a shorter time, and under fully sterile conditions.

3.2 *CANNABIS SATIVA*

3.2.1 Background

C. sativa, also referred to as *Cannabis indica* or *Indian hemp*, is an annual herb. It belongs to the family Cannabaceae. The plant is utilized as food, fibre, and medicine. The plant is native to Central Asia and cultivation is practised especially for medicinal purposes since as early as 900 BC. *C. sativa* is used at the time of religious ceremonies as well.

- The germination of *Cannabis* seeds takes around 12 hours to 8 days.
- Post-germination of 2–4 days, the seed coat splits open and causes the exposure of root and two circular embryonic leaves, cotyledons.
- The seedling phase of the plant lasts 1–4 weeks. This time is considered the greatest vulnerability in the life cycle of *C. sativa*. This time requires moderate humidity levels with sufficient but not excessive soil moisture. The light intensity needed is from medium to high.
- In the vegetative phase, the plant continues to grow and develops new leaves.
- The *C. sativa* plant growth increases significantly in the pre-flowering phase and it develops more branches and nodes.
- The time of flowering can vary from 6 to 22 weeks.

Kingdom	**Plantae**
Subkingdom	Tracheobionta
Superdivision	Spermatophyta
Division	Magnoliophyta
Class	Magnoliopsida
Subclass	Hamamelididae
Order	Urticales
Family	Cannabaceae
Genus	*Cannabis*
Species	*sativa*
Binomial name	*Cannabis sativa*

3.2.2 Ethnomedical Considerations

The use of *Cannabis* in the form of medicine was started before the Christian era in Asia. In India, the drug has a long history and is precisely used as a psychoactive agent and primarily used for various medical and recreational reasons. *C. sativa* was originally a native plant of Western and Central Asia (Figure 3.2). The plant has

FIGURE 3.2 Morphology of *Cannabis sativa*. (A) Leaf, (B) Seeds, and (C) Powder.

been cultivated since ancient times, especially in Asia and Europe. The cultivation in India is controlled and permitted in some specific districts like Almora, Garhwal, Nainital, Kashmir, and Travancore.

3.2.3 Morphology and Pharmacognostic Characteristics

3.2.3.1 Morphology

C. sativa is an annual herb having erected stems. The height of the plant is around 3–10 ft. It is very slightly branched and found to have greyish-green hairs. The leaves of *C. sativa* are palmate, with five to seven leaflets. The leaf has long thin petioles along with acute stipules found at the base of the leaf. Linear-lanceolate, tapering at both ends, is seen in the leaf. The margins of the leaf are sharply serrated. The leaves have many health benefits and are used to reduce excitement, irritation, and pain. The leaf is also used to induce deep sleep. The flowers of *C. sativa* are unisexual. The male flower is available in axillary and terminal panicles, apetalous, and found to have five yellowish petals along with five poricidal stamens. The female flowers of the plant germinate in the axils and terminally and are found to have one single-ovulate ovary. The fruit of *C. sativa* is small and smooth. The colour is light brownish-grey and completely found to be filled by the seed.

3.2.3.2 Microscopic Characteristics

The microscopic study of TS of the root and stem of *C. sativa* shows microscopic features, the cuticle shows the characteristics of dicotyledon. The hairs are present. The presence of regular and radial patterns of the cortex, xylem, and phloem can be seen. The presence of the pericycle can be observed beneath the cortex. The pith is seen in the central region of the TS of the stem. The TS of the root was found to have a regular pattern of the cortex and was seen below the epidermis with a metaxylem.

3.2.4 Phytochemistry

Various phytochemicals of *C. sativa* are identified. Around 483 unique compounds are explored and the following are the five important cannabinoids available in the *Cannabis* plant:

- Cannabidiol
- β-caryophyllene
- Tetrahydrocannabinol
- Cannabigerol
- Cannabinol

Tetrahydrocannabinol is a primary compound that is mainly responsible for the psychoactive effects and it is believed to interact with various brain parts and is normally controlled by the endogenous cannabinoid anandamide.

Cannabidiol is a major constituent of cannabis, and around 40% of extracts consist of cannabidiol. Cannabinol is a therapeutic active cannabinoid and is produced as a metabolite product of tetrahydrocannabinol. It acts as a weak agonist of the CB1 and CB2 receptors and found to have less affinity as compared to tetrahydrocannabinol. β-Caryophyllene has been found to be an elective activator of the CB2 receptor. It is concentrated in cannabis essential oil (12–35%)

3.2.5 Pharmacological Properties

C. sativa L. and its bioactive compounds, including cannabinoids, non-cannabinoids, and cannabidiol, have emerged as a promising intervention for cancer research. The anti-tumour activity of *C. sativa* has been proved in many experimental models of cancer. The evaluation of anti-tumour activity on glioma cells showed that THC, as well as other cannabinoids, induces the apoptotic death of glioma cells. The possible mechanism suggested is by cb1- and cb2-dependent stimulation of the *de novo* synthesis of pro-apoptotic sphingolipid ceramide. Gene expression profiles of the sensitive and resistant glioma cells have given further insight and revealed that cannabinoids are responsible for the activation of specific signalling events downstream of ceramide.

3.2.6 Scientific Investigation for the Management of Cancer

Daris (2019) has described that different plant-derived cannabinoids and cannabis-based pharmaceutical drugs have potential anti-tumour activity, especially in cancer cells that overexpress CB1 and/or CB2 receptors compared to normal tissues. According to the National Cancer Institute, the health benefits of cannabis include anti-cancer properties, antiemetic effects, appetite stimulation, pain relief, and improved sleep.

3.2.7 Reported Mechanism of Action as Anti-cancer Activity

The cannabinoids affect many essential cellular processes and signalling pathways which are crucial for tumour development. The other mechanisms of action of cannabis may include cell cycle arrest, inhibition of cell growth, viability, proliferation, invasion migration and angiogenesis in tumour cells, enhanced apoptosis, and suppression of specific proinflammatory cytokines.

3.2.8 Toxicological Aspects

The plant *Cannabis* has a few toxicological aspects like hallucinogenic, hypnotic and sedative, and anti-inflammatory. The hemp derivatives are suggested for treating glaucoma and as an antiemetic in cancer chemotherapy.

3.2.9 Conclusion and Future Aspects

Hemp is a very special and versatile plant that can produce large amounts of biomass in a very short amount of time. More than 540 phytochemicals have been identified in hemp, and the pharmacological effects of these compounds appear to extend far beyond the psychotic effects of cannabis. Hemp has the potential to address a variety of medical needs, including the alleviation of nausea and anorexia caused by chemotherapy, as well as the symptomatic mitigation of multiple sclerosis. Moreover, it is expected that future studies will discover novel molecular targets of cannabinoids.

3.3 *OCIMUM SANCTUM*

3.3.1 Background

O. sanctum is recognized as Tulsi and designated as the "*Queen of herbs*." The plant is also described as sacred and found to possess many health benefits as mentioned in the ancient literature. *O. sanctum* has a wide distribution, covering almost the entire Indian subcontinent and ascending up to 1,800 m in the Himalayas.

Kingdom	Plantae
Subkingdom	Tracheobionta
Superdivision	Spermatophyta
Division	Magnoliophyta
Class	Magnoliopsida
Subclass	Asteridae
Order	Lamiales
Family	Lamiaceae
Genus	*Ocimum*
Species	*sanctum*

3.3.2 Ethnomedical Considerations

The ethnomedical studies on *O. sanctum* are carried out in order to discover novel compounds for use in indigenous medical systems (Figure 3.3). Tulsi is commonly known for its general vitalizer property and also increases physical endurance. The stem, as well as leaves of holy basil, is found to contain many constituents having an important biological activity, like saponins, flavonoids, triterpenoids, and tannins, etc. that exhibit antioxidant as well as anti-inflammatory activities.

3.3.3 Morphology and Pharmacognostic Characteristics

3.3.3.1 Morphology

The holy basil plant is a small annual or short-lived perennial shrub, up to 1 m (3.3 ft) in height. The stems are found to be hairy and possess simple toothed or maybe entire leaves arranged oppositely along with the stem of the plant. The leaves are

FIGURE 3.3 *Ocimum sanctum*. (A) Tulsi leaf with seeds, (B) Dried powder of Tulsi leaf, and (C) Tulsi leaves.

found to be green or maybe purple, depending on the variety. The flowers are small and purple or may be white, are tubular in shape and possess green or purple sepals, and are borne in terminal spikes. The fruits are nutlets and produce numerous seeds.

3.3.3.2 Microscopic Characteristics

3.3.3.2.1 Leaf

The petiole shows the presence of a cordate outline, having single-layered epidermis consisting of thin-walled and oval cells that have many covering and glandular trichomes. A glandular trichome has two or eight cells per stalk along with a short, sessile head measuring 22-27 mm in diameter. The epidermis is followed by one or two layers of thin-walled, elongated parenchyma cells. Two vascular bundles are situated on the upper and lower surfaces, the middle one is larger than the other two; a xylem surrounds the phloem.

3.3.3.2.2 Midrib

The epidermis, trichomes, and vascular bundles are found to be similar to that of the petiole, except that the cortical layers are reduced at the apical region.

3.3.3.2.3 Lamina

The epidermis and trichomes are found to be similar to that of the petiole. The palisade is single-layered and followed by a closely packed spongy parenchyma layer along with chloroplast and oleo-resin. The palisade ratio is 3.8. The fragments of epidermal cells have an irregular shape. The starch grains have two to five components and the size is around 3–17 μm in diameter. The presence of anomocytic and diacritic stomata on both surfaces is visible and observed to be slightly raised above the level of the epidermis.

3.3.4 Phytochemistry

The phytochemical examination of fresh leaves and stem of *O. sanctum* extract showed the presence of few phenolic compounds, including cirsilineol, isothymusin, apigenin, and rosameric acid, along with appreciable quantities of eugenol. *O. sanctum* leaves have 0.7% volatile oil that consists of around 71% eugenol and about 20% methyl eugenol. Carvacrol and sesquiterpene hydrocarbon caryophyllene can also be found in the oil.

3.3.5 Pharmacological Properties

Tulsi is referred to as "the elixir of life." *O. sanctum* is known to promote longevity. Various parts of plants are used in curing many diseases in Ayurveda and Siddha systems of medicine. The drug is used in order to prevent and cure various diseases, including common cold, cough, influenza, fever, sore throat, asthma, hepatic diseases, snake and scorpion sting, fatigue, skin diseases, arthritis, digestive disorders, diarrhoea, etc. The leaves of *O. sanctum* are scientifically explored as good for nerves as well as sharpen memory. *O. sanctum* chewing helps to cure ulcers as well as infections of the mouth. It is also found to be a better remedy for boosting up the immune system. *O. sanctum* protects many infections caused by viruses, bacteria, fungi, and protozoa.

Scientific investigations during the last several decades have shown that various parts of *O. sanctum*—leaves, stem, root, flowers, and seed—have a plethora of biological and pharmacological activities, including anti-oxidant, anti-inflammatory, antiallergic, immunomodulatory, anticoagulant, antimicrobial, antistress, antiulcer, wound-healing, anticataract, analgesic, antipyretic, antihypertensive, antidiabetic, antifertility, central nervous system depressant, cardioprotective, gastroprotective, hepatoprotective, renoprotective, radioprotective, chemopreventive, and anti-cancer properties.

3.3.6 Scientific Investigation for the Management of Cancer

It is abundantly clear that the ethnomedicinal plant known as *O. sanctum* possesses a huge potential not only for the prevention of a wide variety of human cancers. Alterations in GSH, GSH-related enzymes, such as GST and GSH-Px, and several antioxidant enzymes, such as CAT and SOD, also play a pivotal role in the cancer preventive and therapeutic effects exerted by *O. sanctum*. In addition, *O. sanctum* induces apoptosis and assist in cell cycle arrest and slow tumour growth.

3.3.7 Reported Mechanism of Action as Anti-cancer Activity

Extract of *Ocimum* leaf has the ability to inhibit or reduce the events linked with chemical carcinogenesis. It mainly has the following mechanisms of action:

1. Inducing apoptosis
2. Cytotoxic and antioxidant activity
3. Cell death and viability inhibition
4. Cell cycle arrest
5. Slow tumour growth

3.3.8 Toxicological Aspects

O. sanctum L. or tulsi is traditional as well as clinical, and has proven as a medicinal herb for both its application and efficacy. The ethanol extract of *O. sanctum* leaves seems to be non-toxic as is seen after its acute and subacute oral administrations.

3.3.9 Conclusion and Future Aspects

It is clear that the ethnomedicinal plant *O. sanctum* has enormous potential not only for cancer prevention but also for cancer treatment across a wide range of human cancers. The antioxidant, anti-inflammatory, immunomodulatory, anti-proliferative, proapoptotic, anti-invasive, antiangiogenic, and anti-metastatic properties of *O. sanctum* fractions and pure compounds, as well as their ability to modulate a diverse array of enzymatic activities and signal transduction pathways, could explain the observed chemopreventive and anti-tumour therapeutic efficacy.

3.4 *PRUNUS ARMENIACA*

3.4.1 Background

P. armeniaca, commonly referred to as Apricot or Khubani, belongs to the family Rosaceae. The *P. armeniaca* is an important member of the family Rosaceae. It belongs to the genus *Prunus* which has around 98 species. All the species are found to have significant importance as herbal drugs. *P. armeniaca* belongs to subgenera, namely *Prunophora*, and is the major species of *Prunus* which are commonly found as a *Prunus persica*, *P. armeniaca*, *Prunus salicina*, *Prunus domestica*, *Prunus americana*, *Prunus avium*, *Prunus cerasus*, *Prunus dulcis*, *Prunus ceracifera*, *Prunus behimi*, *Prunus cornuta*, *Prunus cerasoides*, *Prunus mahaleb*, etc.

The plant is edible medicinal plant category and is found native in the regions of Central Asia, Iran, Iraq, Turkey, Pakistan, Syria, and Afghanistan. *P. armeniaca* contains polysaccharides, fatty acids, polyphenols, carotenoids, cyanogenic glycosides, as well as volatile components. *P. armeniaca* is also known as the "moon of the faithful" and the ancient Persians referred to it as the "Egg of the sun."

Kingdom	**Plantae**
Order	Rosales
Family	Rosaceae
Genus	Prunus
Subgenus	*Prunus*
Species	*armeniaca*
Binomial name	*Prunus armeniaca*

3.4.2 Ethnomedical Consideration

P. armeniaca is a deciduous plant of the continental region (Figure 3.4). The plant can tolerate extremely cold temperatures as low as −30°C. Considering India, the plant is cultivated mostly in the North West Hills Region, Himachal Pradesh, Uttar Pradesh, and Jammu and Kashmir. Many types of apricots are locally found in India, including Halman and Rakchaikarpo varieties, and are found especially in Leh–Laddakh area of Jammu and Kashmir state.

3.4.3 Morphology and Pharmacognostic Characteristics

3.4.3.1 Morphology

Apricot is a small tree and is on average 7–10 m tall. The trunk size is around 40 cm in diameter and found to have a dense spreading canopy. The leaves of *P. armeniaca* are ovate and are around 5–9 cm long and 4–8 cm wide. The bases of

FIGURE 3.4 Parts of *Prunus armeniaca*. (A) Apricot fruit, (B) Leaf and fruit seed, and (C) Apricot dried seeds.

the leaves are round and have pointed tips. The margins are finely serrated. The flowers are around 2–4.5 cm in diameter and are found to have five white to pinkish petals. The flowers are produced singly or maybe in pairs in early spring. The fruit is a drupe and found a similarity to small peaches. The fruit is around 1.5–2.5 cm in diameter and the colour is from yellow to orange. The fruit of *P. armeniaca* is often tinged red on the side. The seeds are single and found to be enclosed in a hard, stony shell. The seeds have a grainy, smooth texture, except for three ridges that are running down one side.

3.4.3.2 Microscopic Characteristics

The scanning electron microscopy of pollen grains of *P. armeniaca* shows the availability of pollen grains. Pollen grains are medium in size around 51.32 × 25.51 μm for the elliptical shape and 39.03 × 31.22 μm for the triangular shapes.

3.4.4 Phytochemistry

P. armeniaca is found as a rich source of monosaccharides and polysaccharides, polyphenols, carotenoids, vitamins C and K, niacin, organic acids, phenols, iron, volatile compounds, including esters, norisoprenoids, benzaldehyde, and terpenoids.

3.4.5 Pharmacological Properties

Due to the presence of cyanogenic glycosides (mainly amygdalin) in seeds, it is reported to be used as a medicament for the treatment of cancer. Laetrile, a purported alternative treatment for cancer, has also been extracted from *P. armeniaca* seeds. *P. armeniaca* seed oil has been in use against tumour swellings and ulcers since the seventeenth century.

3.4.6 Scientific Investigation for the Management of Cancer

P. armeniaca seed oil has been used to treat tumours, swellings, and ulcers since the seventeenth century. Amygdalin is a promising choice for the treatment of various cancers. It is also reported in many scientific reports that treating human prostate cancer cells with amygdalin induces programmed cell death and it was concluded that amygdalin offers a valuable option for the treatment of prostate cancers.

3.4.7 Reported Mechanism of Action as Anti-cancer Activity

Apricot anti-cancer mechanisms include lowering cell growth, activating autophagy, triggering apoptosis, protecting tissues/organs from oxidative damage, and reducing inflammation, angiogenesis, and telomerase activity. Its bioactive components inhibited cancer growth through apoptosis, cytotoxicity, angiogenesis, and cell cycle arrest. The bioactive compounds found in *P. armeniaca* possess a dual beneficial effect on oxidative stress in cancer. Firstly, they are responsible for the stimulation of antioxidant defence through the enhancement of antioxidant markers, including SOD, CAT, and GSH, and by decreasing proinflammatory cytokines such as nuclear factor kappa B (NF-κB), tumour necrosis factor (TNF), and interleukins (IL) in tumour mass, it has an antioxidant effect, thus reducing cancer cell growth. NF-κB, TNF, IL, matrix metalloproteinase (MMP), superoxide dismutase (SOD), catalase (CAT), glutathione (GSH), malondialdehyde (MDA), nitric oxide (NO), poly-ADP ribose polymerase (PARP), and VEGF (vascular endothelial growth factor).

3.4.8 Toxicological Aspects

P. armeniaca kernels may also lower blood pressure and hence interact with blood-pressure-lowering herbs and vitamins. Fresh *P. armeniaca* should not be consumed by individuals who have gastrointestinal ulcers, gastritis with excessive acidity, pancreatitis, or liver diseases. Because of the high sugar content, the fruit is also contraindicated in cases of diabetes. Consuming significant amounts of these fruits on a daily basis can result in diarrhoea.

3.4.9 Conclusion and Future Prospects

More studies should be conducted to elucidate the molecular mechanism of interaction of various parts of these plant-based drugs with the human body in different diseases. *P. armeniaca* is a worldwide deciduous plant that has manifold uses.

Extensive research and development work should be undertaken on *P. armeniaca* and its products for their better economic and therapeutic utilization.

3.5 *ZINGIBER OFFICINALE*

3.5.1 Background

Z. officinale, referred to as "Ginger," is considered the most widely consumed spice in India. The plant originates from Southeast Asia and is spread to Europe. Ginger plants have a long history of use. In herbal medicine, *Z. officinale* is used in order to treat many diseases like vomiting, pain, indigestion, cold, etc. The spice ginger is found as an underground rhizome and referred to botanically as *Z. officinale*.

Kingdom	**Plantae**
Order	Zingiberales
Family	Zingiberaceae
Genus	*Zingiber*
Species	*Z. Officinale*
Binomial name	*Zingiber officinale*

3.5.2 Ethnomedical Considerations

The family of *Z. officinale* is Zingiberaceae and possesses lots of medicinal, nutritional, and ethnomedical values. *Z. officinale* is used throughout the world as a spice for its flavouring property and as a herbal remedy, it is found to be used in many systems of medicines, including Ayurveda, Siddha, Arabian, African, and Chinese.

It mainly contains gingerols, zingibain, starch, essential oil, including zingiberene, zingiberol, borneol, etc., bisabolene, oleoresins, mucilage, and proteins (Figure 3.5).

3.5.3 Morphology and Pharmacognostic Characteristics

3.5.3.1 Morphology

Z. officinale is a tuberous perennial root. The size is around more than 1 in. in length. The roots of the plant are flattened on the upper surfaces. The plant is irregularly branched and found to have light ash colour. *Z. officinale* produces an annual leafy stem of about 2–3 ft in height. The leaves of the plant are lanceolate, oblong, and smooth. The length of the leaves is around 5–6 in. and are found to grow alternately along the length of the stem. The colour of the flowers of *Z. officinale* ranges from yellow to purple and yellow-spotted. The flowers have green bracts along with yellow margins.

3.5.3.2 Macroscopic Characteristics

The powdered form of dried ginger is a pale yellow to cream in colour with a pleasant, aromatic odour and a characteristic and pungent taste. The diagnostic characteristics are abundant starch granules, fibres, vessels, oleo-resin cells, and abundant parenchyma.

FIGURE 3.5 Morphology of ginger.

3.5.3.3 Microscopic Characteristics of Ginger Rhizome

It includes cork which is absent in Jamaica ginger, phellogen, cortex and oleoresins, endodermis, and stele. The remaining tissue contains fibrovascular bundles, starch, and oleoresin cells similar to the cortex.

3.5.4 Phytochemistry

Z. officinale was found to have volatile oils (1–2%) bisabolene, citral, citronellal, geranial, gingerol, camphene, linalool, borneol, phellandrene, cineole, and zingiberene. The pungency of *Z. officinale* (ginger) is because of the presence of gingerol (5–8%) and its aroma is due to volatile oils bisabolene, zingiberene, and zingiberol.

Z. officinale consists of an oily liquid that contains phenols referred to as gingerol. Gingerol is mainly responsible for pungency along with the pharmacological activity of ginger. The dehydration of gingerol results in the formation of shogaols, which further degrade in order to produce zingerone. The aroma of *Z. officinale* is because of ginger oil, which contains a mixture of more than 50 constituents. Volatile oils present in ginger oil are mainly sesquiterpenes, including β-bisabolene and zingiberene; monoterpenes, including β-phellandrene, borneol, etc.; and sesquiterpene alcohol as zingiberol.

3.5.5 Pharmacological Properties

Z. officinale is used as antiemetic. The aromatic and carminative property of ginger helps to improve the effects of motion sickness and it acts directly in the GI tract. The plant is found to be used as an antioxidant, antitoxic, eicosanoid balance, enzyme activity, probiotic support, serotonergic, and systemic stimulant. Its demonstrated effects are analgesic, antibacterial, antidiabetic, antiemetic, antifungal, anthelmintic, anti-inflammatory, antithrombotic, anti-tumour, antitussive, antiulcer, antiviral, flatulence, immune supportive, migraine headache, morning sickness, nausea, thermoregulatory, etc.

3.5.6 Scientific Investigation for the Management of Cancer

Ginger contains diverse bioactive compounds, such as gingerols, shogaols, and paradols, and possesses multiple bioactivities, such as antioxidant, anti-inflammatory, and antimicrobial properties. Additionally, ginger has the potential to be the ingredient for functional foods or nutriceuticals, and ginger could be available for the management and prevention of several diseases such as cancer, cardiovascular diseases, diabetes mellitus, obesity, neurodegenerative diseases, nausea, emesis, and respiratory disorders.

3.5.7 Reported Mechanism of Action as Anti-cancer Activity

Breast cancer: Induction of typical apoptotic changes in nuclear morphology, chromatin condensation and fragmentation, membrane shrinkage and blebbing; enabled autophagy followed by caspase-independent apoptosis; induction of autophagy

Prostate cancer: Cell cycle arrest in the G1 phase, followed by a reduction in S and G2/M via a p21-dependent mechanism; downregulation of MRP1 and GST-protein expression

Ovarian cancer: Decreased production of NF-κB-regulated angiogenic factors; p53-induced apoptosis via Bcl-2 deletion

Colon cancer: Cell cycle arrest at several checkpoints by inhibiting cyclin-dependent kinases and activating cell cycle checkpoints.

3.5.8 Toxicological Aspects

Ginger is generally considered a safe herbal medicine. The acute toxicity studies indicate that ginger dosages to elicit acute toxicity are high and higher than usually administered doses of ginger root. There is insufficient evidence that ginger root may cause testicular weight changes. The genotoxicity of ginger root has not been studied. Ginger root has mutagenic as well as antimutagenic properties in microbial test systems.

3.5.9 Conclusion and Future Aspects

Ginger derivatives have been found to possess chemopreventive properties like cell cycle arrest along with increased cellular death (apoptosis, autophagy, and autosis), and redox homeostasis disruption. They also block angiogenesis, the development of

CSCs, and the EMT process. As a result, this natural compound influences tumour cell survival both directly and indirectly, inhibiting invasion and metastasis processes while having no significant toxic effects on normal cells.

3.6 *CURCUMA LONGA*

3.6.1 Background

C. longa is also known as Indian saffron because it has a brilliant yellow colour. Its tuberous rhizome is used as a spice or condiment and in cosmetics. It is a perennial herbaceous plant that belongs to the Zingiberaceae family. It is mainly found in tropical and subtropical regions of the world (China, India, Nepal, Madagascar, Pakistan, Malaysia, Vietnam, etc.). *C. longa* contains curcuminoids (a mixture of DMC [demethoxycurcumin], curcumin, and bisdemethoxycurcumin). Curcuminoids are mainly responsible for the therapeutic activity of the *C. longa*. This rhizome has shown a lot of medicinal uses.

Kingdom	**Plantae**
Order	Zingiberales
Family	Zingiberaceae
Genus	*Curcuma*
Species	*longa*

3.6.2 Ethnomedical Considerations

In Ayurveda, the poultice of *C. longa* is used in the treatment of eye and skin infections, burns, cuts, and healing of wounds. Boiled milk with *C. longa* has been given to cure cough and cold (Figure 3.6). *C. longa* is also utilized in the treatment of different types of gastrointestinal disorders such as ulcers, flatulence, acidity, indigestion, and liver disorder. Curcumin is the main active constituent of it which is utilized for yellow colouring and providing flavours to different types of dishes. In northern India, it is also given to women (after delivery) as a tonic and its poultice is applied to heal the birth canal.

3.6.3 Morphology and Pharmacognostic Characteristics

3.6.3.1 Morphology

Herb: *C. longa* is a perennial herbaceous plant (3 ft and 3 in. plant) and an underground rhizome.

Leaves: The leaves are green in colour (the upper surface is dark green and the lower surface is pale green). It is 1 m long, petiole, with sheath of the same length and lanceolate. Their leaves emerged from the rhizomes.

Rhizomes: The rhizomes are underground stems and their older rhizomes are scaly and brown in colour. Its young rhizomes are yellowish brown in colour (internal surface is bright orange in colour and after cutting yellowish is internal flesh). It is bitter in taste and it has characteristic odour.

Flowers: The flowers are white with a deep ferruginous purplish colour. It has a tall spike. It has a pseudostem and is 2 m long.

FIGURE 3.6 Photographs of *Curcuma longa*. (A) Herb of *Curcuma longa*, (B) Rhizome of *Curcuma longa*, and (C) Flower of *Curcuma longa*.

3.6.3.2 Microscopic Characteristics

Parenchyma cell: This has rectangular and slightly irregular parenchymatous cork (four to six layers) cells. It has gelatinized starch (yellow clumps).

Trichomes: These have a quite distinct, unicellular, elongated, and conical shape. It is bluntly pointed with a thick wall (moderate).

Cortex: Thin-walled rounded parenchymatous cells containing scattered vascular bundles (collateral). Prismatic and cluster types of calcium oxalate crystals are present.

Ground tissue: It is brown in colour due to the presence of oleoresin cells throughout.

Oil cells: Have suberized cell walls.

Pith: Vascular bundles (scattered) forming discontinuous rings under endodermis.

Endodermis: Starch grains are abundant.

3.6.4 Phytochemistry

C. longa contains a lot of classes of active phytoconstituents like alkaloids, polyphenolics, flavonoids, carbohydrates, proteins, saponin, tannins, and terpenoids. *C. longa* is a rich source of curcuminoids (polyphenolic) such as curcumin (Curcumin I [diferuloylmethane, ~77%]), Curcumin II (caffeoyl feruloyl methane, ~17%) and Curcumin III (dicaffeoylmethane, ~3%). Curcumin is yellow in colour and is utilized as a food colouring agent. Curcumin is found in keto and enol tautomeric forms. Curcumin is melted at 176–177°C. Curcumin is soluble in ethanol, acetic acid, alkaline, etc. and insoluble in water. Some other phytoconstituents are also found such as fat, minerals, moisture, essential oil, sabinene, cineol, borneol, α-phellandrene, zingiberene, and sesquiterpenes.

3.6.5 Pharmacological Effects

Curcumin is a main phytopolyphenol pigment that is isolated from the *C. longa* Linn. It has a lot of therapeutic activities. It has shown antiseptic, antiviral, antibacterial, antimicrobial, anti-inflammatory, antioxidant, anti-tumour, antiandrogenic, wound-healing, and immunostimulant properties.

3.6.6 Scientific Investigation for the Management of Cancer

Curcumin inhibits the proliferation of cancer cell lines such as prostate carcinoma PC-3 cells, MDA-MB-231 breast cancer cells MCF-7 cells, HCT-8/VCR cells, HCT-15 cells, and HepG2 cells. Curcumin has been shown in clinical trials to be useful in decreasing and preventing cancers such as multiple myeloma, colon, pancreas, breast, prostate, and lung cancer. *In vitro* and *in vivo* studies on colon cancer cells using a monocarbonyl analog of B63 obtained through some chemical modifications of curcumin's structure revealed that this component has an antiproliferative effect while also suppressing tumour growth.

3.6.7 Reported Mechanism of Action as Anti-cancer Activity

Curcumin's key modes of action include inducing apoptosis and reducing tumour proliferation and invasion through the suppression of a range of cellular signalling pathways. Curcumin has been shown in several trials to have anti-tumour activity against breast cancer, lung cancer, head and neck squamous cell carcinoma, prostate cancer, and brain tumours, demonstrating its potential to target numerous cancer cell lines. Despite all of the advantages listed above, curcumin is utilized less due to its poor water solubility which results in low bioavailability (oral) and chemical stability. Another barrier is curcumin's limited cellular absorption. Because curcumin is hydrophobic, it tends to permeate the cell membrane and attach to the fatty acyl chains of membrane lipids via hydrogen bonding and hydrophobic interactions, resulting in low curcumin availability inside the cytoplasm.

3.6.8 Toxicological Aspects

Curcumin ameliorates chemotherapy-induced gastrointestinal toxicity, reduces chemotherapy-induced cardiotoxicity, prevents chemotherapy-induced hepatotoxicity, ameliorates chemotherapy-induced nephrotoxicity, decreases chemotherapy-induced ototoxicity, attenuates chemotherapy-induced myelosuppression, attenuates chemotherapy-induced neurotoxicity, and prevents chemotherapy-induced genotoxicity.

3.6.9 Conclusion

C. longa is regarded as a treasure trove of medicinal properties and has been used in most traditional systems of medicine, particularly the Unani system of medicine. Curcumin is a safe natural product with a lower cost than many drugs, which suggests that it may be effective in the prevention and treatment of a variety of disorders.

3.7 *Smilax zeylanicum*

3.7.1 Background

The common name of *S. zeylanicum* L. is chopachinee in Sanskrit and chobchini in Hindi. It is a deciduous climbing shrub and its stem is substituted with Sarsaparilla. It comes under monocotyledon plants and belongs to the Smilacaceae family. Its genus *Simplex* Linn. has 350 species in the whole world. In India, it has 24 species. In South India, its genus has four species: *S. zeylanica* Linn., *S. aspera* Linn., *S. perfoliata* Roxb., and *S. wightii* A. It is found in tropical and subtropical areas of the northern Himalayan region to southern Kerala (altitude 500–1800 m). It contains a lot of medicinal active phytoconstituents like alkaloids, flavonoids, tannins, glycosides, and phenylpropanoid glucosides.

Kingdom	**Plantae**
Order	Liliales
Family	Smilacaceae
Genus	*Smilax*
Species	*zeylanicum*
Binomial name	*Smilax zeylanicum*

3.7.2 Ethnomedical Considerations

Various parts, viz. the tuber, stem, root, and leaves, of *Simplex zeylanicum* are utilized for various reasons (including fruit, food, medicine, and dye) by the Ayurvedic system and many tribal areas of the world (Figure 3.7). Its different parts (root, stem, leaves, tubers, etc.) are used either singly or in combination in certain formulations such as paste and decoction. These formulations are used throughout the world to cure and treat several human and veterinary ailments. Its parts (root, rhizomes, and leaves) are used for the treatment of fever, epilepsy, soreness, swelling, abscesses, and skin disease. Its roots are mainly used for the treatment of rheumatism and lower body pain. It is also used in bloodless dysentery and healing techniques.

FIGURE 3.7 Photographs of *Smilax zeylanica.* (A) Fruit of *Smilax zeylanica*, (B) Leaves of *Smilax zeylanica*, and (C) Flower of *Smilax zeylanica.*

In the villages of Bangladesh, it is used for the treatment of fever, wounds, and headache.

3.7.3 Morphology and Pharmacognostic Characteristics

Smilax zeylanica is a woody climbing large shrub, dioecious, and has prickles and angular branches. It comes under the monocotyledon plants and belongs to the Smilacaceae family. It has 350 species in the whole world. It is found in tropical and subtropical areas of the northern Himalayan region to southern Kerala (altitude 500–1,800 m).

Leaves: Simple, alternate (20 × 11 cm), lanceolate, coriaceous, opposite acute cuspidate at the apex. Its base is narrow and has reticulate venation. It has a petiole (3 cm long), and tendril on either side above the base. Its leaves are green in colour, shining on the upper side. It has an aromatic odour and is bitter in taste

Flowers: These are greenish-white in colour, dioecious (six stamens in male flowers and usually three staminodes in female flowers) and umbellate (one to three umbels on axillary peduncle). Its anther is introrse. Pistillode is absent. Flowering happens in the months of November and April.

Fruit: It is globose, smooth, leathery, fleshy berries and after ripening, it is converted into red. It has one to two seeds.

3.7.3.1 Microscopic Characteristics

Lamina (250 μm): The upper epidermis is thick, almost rectangular. The lower epidermal remains thin, and has spindle-shaped small cells. The mesophyll tissue consists of the upper part of two layers of small, straight, oblong, compact cells and an abaxial part of five or six much lobed spongy parenchymatous cells which are joined with each other forming wide air spaces. Calcium oxalate druces are seen scattered in the surface view of the lamina. Druces occur within modified circular lithocysts. The druces are 60 μm in diameter.

Leaf margin: The leaf margin (170 μm thick) is conical with a blunt end. The epidermal layer is slightly large (thicker cuticle). The extreme margin of the lamina consists of compact thick-walled cells. The submarginal part has normal palisade spongy differentiation of the mesophyll.

3.7.4 Phytochemistry

The *S. zeylanica* contains active phytoconstituents such as 1–3% steroidal saponins, alkaloids, phytophytosterols, starch, resin, flavonoids, sarsapic acid, and minerals. Its leaves contain diosgenin, sterol, flavoids, saponin hydroxytyrosol, squalene, *trans*-isoieugenol, tannins, and triterpenoids. Its roots and rhizomes contain diosgenin, phytosterol, tannins, steroids, flavonoids, phenols, and saponin. Its stem contains steroids, tannins, flavonoids, and glycosides. *S. zeylanica* also contains some nutrients like nitrogen, phosphorus, potassium, sodium, calcium, magnesium, copper, iron, manganese, and zinc.

3.7.5 Pharmacological Properties

Smilax zeylanica is a medicinal potential plant which has significant pharmacological effects and anti-cancer activities.

Antioxidant activity: Its roots and rhizomes have shown antioxidant activity. Methanolic extract of *S. zeylanica* has showed potential scavenging effect against DPPH (2,2-diphenyl-1-picryl-hydrazyl-hydrate). The ethanolic extract of the stem showed maximum DPPH (2, 2-diphenyl-1-picryl-hydrazyl-hydrate) scavenging activity. This is concluded by the *in vitro* propagation and free radical studies of *S. zeylanica* by multiple shoots formation from nodal segments.

Anti-cancer activity: Rhizome extract of *Smilax china* has shown anti-cancer activity against HeLa cells. It is assessed by MTT assay and clonogenic assay. Its extract contains flavonoids (kaempferol-7-*O*-β-D-glucoside) which are responsible for the activity.

3.7.6 Scientific Investigation for the Management of Cancer

Plant-derived phytochemicals have phenols and flavonoids which show antioxidant properties that act as a tool to scavenge reactive oxygen species. Studies are done to analyse *in vitro* antioxidant and cytotoxic potential of methanolic extract and petroleum ether extracts of *S. zeylanica* L. stems. Studies have shown that methanolic extract is a potential source of natural antioxidants, whereas petroleum ether extracts have potential for promising anti-cancer molecules.

3.7.7 Reported Mechanism of Action as Anti-cancer Activity

Studies shows that the potential chemoprevention of methanol extract of *S. zeylanica* leaves might be due to stabilization and increase in all the components of the antioxidant system which is due to antioxidant and free radical scavenging activity. *S. zeylanica* leaf extract decreases the extent of lipid peroxidation with concomitant increase in the activities of enzymatic antioxidants (superoxide dismutase, catalase, glutathione peroxidase, glutathione reductase, and glutathione-*S*-transferase) and non-enzymatic antioxidants (reduced glutathione, vitamin C, and vitamin E) levels.

3.7.8 Toxicological Aspects

The acute toxicological studies on *S. zeylanica* have revealed that the methanolic extract of the herb is safe and can be used efficiently as an anti-cancer agent.

3.7.9 Conclusion and Future Aspects

Recent enhancements in research made it possible for providing scientific reasons for the traditional uses of this plant. Many phytochemicals are studied like smilagen, diosgenin, sitosterols, hydroxytyrosol, sarsasapogenin, phenol, alkaloids, etc. Various bioactivities have been proven such as antibacterial, thrombolytic, antioxidant, antidepressant, hepatoprotective, analgesic, antidiabetic activities.

It can be concluded that *S. zeylanica* is an ethnobotanical treasure and has a wide range of bioactivities. These facts make it a highly promising candidate for making natural drugs. Moreover, further bioactive phytochemical screening of *S. zeylanica* can give a new approach for developing holistic medicine.

3.8 *TINOSPORA CORDIFOLIA*

3.8.1 Background

T. cordifolia is commonly known as Indian bitter (English), Giloya (Hindi), Amrita (Sanskrit), Amruthaballi (Kannada), and Guduchi. It belongs to the family Menispermaceae (it consists of 70 genera and 450 species). It is a shining (glabrous), deciduous, climbing shrub (woody) having greenish-yellow flowers. It is mainly found in tropical regions at higher altitudes. It is distributed all over India, Sri Lanka, China, Africa, etc. According to Ayurveda, it has immunomodulatory activity and it increases the defence against precise microorganisms. According to the Indian system of medicine, it has a lot of medicinal properties, so it is used in various types of diseases like cough, cold, gastrointestinal diseases, fever, jaundice, urinary tract infection, skin disease, chronic diarrhoea, arthritis, etc.

Kingdom	**Plantae**
Division	Magnoliophyta
Class	Magnoliopsida
Order	Ranunculales
Family	Menispermaceae
Genus	Tinospora
Species	*T. cordifolia*

3.8.2 Ethnomedical Considerations

T. cordifolia is used as folk medicine in different tribes and their areas (Figure 3.8). The root paste of *T. cordifolia* and *Solanum surattense* is used in fever in Baiga tribes which are present in Uttar Pradesh. The whole plant extract of *T. cordifolia* is used in diarrhoea, jaundice, and fever in the tribals of Maharashtra. The whole plant extract of *T. cordifolia* is used in a bone fracture in different Muslim tribes of Rajouri, Tawi, Gujjars, and Backwals.

FIGURE 3.8 Photographs of *Tinospora cordiolia.* (A) Leaves of *Tinospora cordiolia*, (B) Stem of *Tinospora cordiolia*, (C) Flowers of *Tinospora cordiolia*, (D) Fruits of *Tinospora cordiolia*, and (E) Seeds of *Tinospora cordiolia.*

3.8.3 Morphology and Pharmacognostic Characteristics

3.8.3.1 Morphology

T. cordifolia is a coiling climber, a succulent, glabrous shrub that is found in tropical and subtropical regions of India. Its different parts have the following morphological characteristics:

Leaves: These are simple (alternate), petiolate (long), chordate-shaped (heart-shaped), reticulate venation, pulvinate, and lamina is ovate.

Stems: These are grey and creamy white, filiform, plumpy, lenticels (rosette), coiling, and longitudinal branches. It has a characteristic odour and bitter taste.

Flowers: These are greenish-yellow in colour. It is small, recemes, and unisexual. Its male flowers are clustered and female flowers show solitary inflorescence. Six sepals are arranged in two whorls. They are ovate and membranous.

Fruits: These are orange-red in colour, plumpy (smooth and fleshy), and oval-shaped.

Seeds: These are white in colour. They are found in bean and curved shapes.

3.8.4 Phytochemistry

T. cordifolia contains different classes of active phytoconstituents such as alkaloids, flavonoids, terpenoids, phenolics, polysaccharides, phenolics, polysaccharides, sesquiterpenoids, glycosides, and steroid triterpenoids. Its leaves are also a good source of Ca, P, and proteins. The major active phytoconstituents in the stems of *T. cordifolia* are tembertarine, tinosporin, columbin, tinosporide, tinosporaside, tinosporidine, cordifolide, palmatine, cordifol, berberine, heptacosanol, β-sitosterol, clerodane furano diterpene, magniflorine, diterpenoid furano lactone, choline, and tinosporin. Tinocordifolin (new daucane-type sesquiterpene) is isolated from the stems of *T. cordifolia*. The four amritosides A, B, C, and D and their acetate derivative clerodane furano diterpene glucoside have been isolated from this plant. Their structure was established by spectroscopic analysis. Steroids such as makisterone A, β and δ-sitosterol, ecdysterone, 20-β-hydroxyecdysone, and giloinsterol were isolated from its aerial and stem parts. Some alkaloids such as palmatine, berberine, tembetarine, magnoflorine, choline, tinosporin, and isocolumbin palmatine were isolated from the stem and root parts of *T. cordifolia*.

3.8.5 Pharmacological Properties

In the Indian system of medicine, *T. cordifolia* is a medicinal valuable plant that has a lot of medicinal properties, so it is used as a general tonic, antiallergic, antidiabetic, antispasmodic, antipyretic, wound-healing, hepato-protective, antileprotic, immunomodulators, anti-cancer. It is used in asthma and pneumonia. It also has cholesterol-lowering property.

The stems of *T. cordifolia* are used as a diuretic and bitter tonic, to cure jaundice and to treat skin problems. The root and stem of *T. cordifolia* are used with a combination of other drugs as an antidote to snake bites.

T. cordifolia is used to cure the side effects which are caused by chemotherapy and radiation therapy during the treatment of cancer. It is also used to heal foot ulcers that are caused by diabetes.

T. cordifolia contains potential active phytoconstituents which have shown significant pharmacological activities.

3.8.6 Scientific Investigation for the Management of Cancer

Giloe, *T. cordifolia*, and some of its active phytochemicals have been reported to trigger cytotoxic effects in various cultured human cancer cells and also some animal tumours *in vivo*. The plant's products may be useful due to the presence of numerous chemical molecules in them, which may act on cancer cells through multiple mechanisms, making tumours more amenable to treatment with less or no adverse side effects, unlike chemotherapy.

3.8.7 Reported Mechanism of Action as Anti-cancer Activity

Giloe and its phytochemicals like palmatine and berberine have been reported to be cytotoxic in various cultured human cell lines and also in some preclinical transplanted tumour models. The cytotoxic effect of giloe and its constituents may be due to its ability to stimulate the free radical formation and DNA damage in the tumour cells. They also reduce the antioxidant status of tumour cells by increasing lipid peroxidation and lactate dehydrogenase. Topoisomerases trigger DNA damage which causes cytotoxic effects in tumour cells. *T. cordifolia* may have inhibited topoisomerases at the molecular level. Cytotoxicity in tumour cells may get induced when NF-κB, COX-II, Nrf2, STAT3, of Bcl-2, $Ca^{2}+$ release, cyclin-dependent kinase (CDK) 2, CDK4, and cyclin (B, D, and E) are suppressed. The activation of p53, Weel, and CDk1 proteins, Bax, P27, procaspase-9, caspase-9, caspase-3, and poly (ADP-ribose) polymerase (PARP) would have led to increase in apoptotic death of tumour cells.

3.8.8 Toxicological Aspects

T. cordifolia is safe and does not produce harmful effects. A dose of 500 mg/day of *T. cordifolia* is given for 21 days to healthy individuals in a clinical study. It has not shown adverse effects on the cardiovascular system (CVS), renal system, central nervous system, and gastrointestinal system.

3.8.9 Conclusion and Future Aspects

The pharmacological actions of *T. cordifolia* (as per Ayurvedic texts) are validated which suggests that this drug has enormous potential in the pharmacotherapeutic field. Future studies can be performed to explore its molecular mechanisms.

3.9 *EMBLICA OFFICINALIS*

3.9.1 Background

E. officinalis is also known as amla, *Phyllanthus emblica*, and Indian gooseberry. It belongs to the Euphorbiaceae family. It is found in tropical and subtropical regions in various countries like China, Indonesia, India, and Southeast Asia.

E. officinalis has its vernacular names in different language like Sanskrit (dhatriphala, sriphalam, amla, and vayastha), Hindi (amla), English (*Emblica myroblan*), Italian (*Mirabolano emblico*), and French (*Phyllanthe emblica*). It is a magical medicinal plant that is a valuable gift by nature to humans for its health benefits. According to ancient Indian historical methodology, it is the first holy plant which is created by the universe. It is mainly found in native India. It is found in tropical and subtropical countries like Pakistan, China, South East Asia, Sri Lanka, and Malaysia.

Kingdom	**Plantae**
Phylum	Magnoliophyta
Class	Magnoliopsida
Order	Euphorbiales
Family	Euphorbiaceae
Genus	Emblica
Species	*E. officinalis*
Binomial name	*E. officinalis* L.

3.9.2 Ethnomedical Considerations

E. officinalis is used in different traditional systems of medicine; according to Ayurveda, different parts of *E. officinalis* have significant medicinal properties (Figure 3.9). Preclinical studies have revealed that *E. officinalis* has several properties like cardioprotective, antiatherosclerotic, gastroprotective, adaptogenic, hepatoprotective, antipyretic, analgesic, nephroprotective, antitussive, antiatherogenic, antianaemia, antihypercholesterolemia, wound-healing, antidiarrhoeal, and neuroprotective properties. Experiments have revealed that *E. officinalis* and its phytochemicals (gallic acid, ellagic acid, pyrogallol, corilagin, geraniin, elaeocarpus, and prodelphinidins Bl and B2) possess antineoplastic effects. *E. officinalis* is revealed to have radio-modulator, chemomodulator, and chemopreventive effects. It has also shown free radical scavenging, antioxidant, anti-inflammatory, antimutagenic, and immunomodulatory activities. These properties are beneficial for the treatment and prevention of cancer

3.9.3 Morphology and Pharmacognostic Characteristics

Plant: It is an 8–18 m high deciduous tree. It has shiny grey bark exfoliating in unusual small flakes.

Leaves: These are simple (pinnate), sub-sessile (closely branchets), and light green in colour.

FIGURE 3.9 Photographs of *Emblica officinalis*. (A) Plant of *Emblica officinalis*, (B) Leaves of *Emblica officinalis*, (C) Fruits of *Emblica officinalis*, (D) Fruits of *Emblica officinalis*, and (E) Flowers of *Emblica officinalis*.

Fruits: These are spherical, globular shine, and green in colour (unripe fruit). The ripe fruit gets converted into light brown colour (its mesocarp is yellow in colour and endocarp is yellowish brown in colour). It has 15–20 mm length and 18–25 mm width and its weight is 60–70 g. It has a small and slight conic depression on both apexes.

Seeds: These are 4–6 in number, smooth, and dark brown in colour.

Flowers: These are greenish-yellow in colour.

Barks: The thickness of the bark is up to 12 mm and it is shiny greyish brown or greyish in colour.

3.9.3.1 Microscopic Characteristics

Fresh fruit of *E. officinalis* has shown single-layered epidermal cells. They are tubular in shape and are covered with cuticle (thick) cells. Beneath the two- to four-layered hypodermis (elongated), it possesses thick cell walls. Epidermal cells are bigger than hypodermal cells. The bulk of fresh fruit of *E. officinalis* consists of mesocarp cells and parenchymatous cells (thin-walled). Parenchymatous cells of the mesocarp contain vascular bundles which are found in the form of a vascular network. Vascular bundles are variable in size and shape. Each vascular bundle is made up of xylem (thick-walled) and phloem elements (consisting of sieve and companion cells).

The coarse powder of *E. officinalis* is greyish-white, dark brownish, and black in colour. In the microscopic study, it has shown lignified tissues (brown in colour), aleurone grains (green to brown colour), and prismatic crystals of silica (brown colour).

3.9.4 Phytochemistry

Fruits: The fruits of *E. officinalis* contain minerals (Ca, P, and Fe), fibres, carbohydrates, protein, fat, nicotin vitamins (thiamine, niacin, and riboflavin), vitamin C (rich source), tannin (emblicanin A and B), nicotionic acid, phyllem-belin, phyllem-acid, putranjivain A, gallic acid, ellagic acid, and pectin.

Seeds: The seeds of the *E. officinalis* contain a fixed oil (brown in colour), phosphatides, fatty acids (linolenic, linoleic, oleic, stearic, palmitic, and myristic acid), and essential oil.

Leaves: The leaves and bark of *E. officinalis* are the major sources of tannins.

The leaves and bark are rich in tannins. The leaves contain active phytoconstituents like gallic acid, ellagic acid, chebulic, chebulagic, chebulinic acids, gallotannin (amlie acid), alkaloids phyllantidine, and phyllantine.

3.9.5 Pharmacological Properties

E. officinalis possess bioactive phytoconstituents: tannins, flavonoids, saponins, terpenoids, ascorbic acids, etc. These phytoconstituents have diverse pharmacological activities such as antimicrobial, antioxidant, anti-inflammatory, radio-protective, hepatoprotective, immunomodulatory, hypolipidemic, etc. *E. officinalis* is also reported to have antidepressant, anti-cancer, antidiabetic, anti-HIV-reverse transcriptase, antiulcerogenic, and wound-healing activities.

3.9.6 Scientific Investigation for the Management of Cancer

E. officinalis has been used in traditional medicine for generations to treat symptoms ranging from constipation to the treatment of tumours. However, the potential of *E. officinalis* extract can be used as an anti-cancer agent which has been proven (using medical techniques) to contain molecules with both cancer-preventative and anti-tumour activities.

3.9.7 Reported Mechanism of Action as Anti-cancer Activity

E. officinalis exhibits its anti-cancer activities through inhibition of AP-1 and targets transcription of viral oncogenes responsible for development and progression of cervical cancer, thus indicating its possible utility for treatment of HPV-induced cervical cancers.

There are four possibilities. Firstly, *E. officinalis* has potent free radical scavenging activities that might prevent reactive oxygen species–induced DNA damage and oncogenesis. Secondly, the extract has properties allowing it to reduce the levels of cytochrome enzymes in liver cells. Thirdly, the extract has anti-inflammatory activities that might prevent inflammation-related cancers. Finally, the flavonoid quercetin has been demonstrated to attenuate tumour growth in multiple animal models.

3.9.8 Toxicological Aspects

Exposure to low levels of extract of these berries may impair tumour progression (at early stages). There is an issue regarding potential hepatotoxicity after long-term ingestion of *E. officinalis*. Before *E. officinalis* extract can be safely recommended for long-term consumption for the prevention of cancer, this issue of hepatotoxicity is to be resolved in the future by clinical and epidemiological studies.

3.9.9 Conclusion and Future Aspects

For *E. officinalis* to become relevant clinically, it is imperative that the molecules mediating the anti-tumour effects of the plant be identified and even more potent, patentable derivatives synthesized. Without the possibility of patents, the pharmaceutical industry will undoubtedly not invest the enormous amount of money required to carry out clinical trials using these putative chemotherapeutics. Such evidenced-based trials will eventually be necessary to prove the worth of these extracts in preventing and treating human cancer.

3.10 *WITHANIA SOMNIFERA*

3.10.1 Background

Withania belongs to the family Solanaceae and is widely known as an immune-modulating agent from ancient times (Figure 3.10). The traditional Indian literature such as Ayurveda, Siddha, and Unani system revealed a variety of therapeutic benefits of Ashwagandha. It is also known as "winter cherry" and "Indian ginseng."

In Indian pharmacopoeia, it is reflected as an official drug for the treatment of every ailment. The useful part in Ashwagandha is roots and leaves that are considered as tonic, aphrodisiac, in dyspepsia, rheumatism, and weakness. These plants are also useful in the treatment of syphilis and decoction of roots is advised in asthma.

Kingdom	**Plantae**
Order	Solanales
Family	Solanaceae
Genus	*Withania*
Species	*W. somnifera*

3.10.2 Ethnomedical Considerations

The Ashwagandha plant has been utilized for numerous pharmacological actions and as a main ingredient of geriatric tonics. In Ayurveda, it is proclaimed as a potent aphrodisiac rejuvenation and helps in maintaining life's expectancy. It is beneficial in stress management, and has been proved as a potential agent for cardiovascular and neurodegenerative disorders. Ashwagandha roots are traditionally claimed for treatment of brain disorders, insomnia, skin problems, coughing, tiredness, nervous exhaustion, impotency, debility, convalescence, dehydration, bone weakness,

FIGURE 3.10 Plant of *Withania somnifera*.

loosening of teeth, thirst, premature ageing, and muscle tension. It is employed in boosting energy, fortitude, slower ageing process, and strength and fostering the essential elements and fluids of the body such as increased muscle fat, lymph, semen, blood, and cell proliferation. It helps to potentiate learning ability and memory capability. The active therapeutic component withanolide also acts as antifungal, antibacterial, and anticarcinogenic activities.

3.10.3 Morphology and Pharmacognostic Characteristics

The *Indian ginseng* or *Radix Withaniae* or winter cherry obtained from the dried roots of *W. somnifera* (L.) Dunal. belongs to the Solanaceae family. The genus *Withania* comprises 23 species, including *W. somnifera* L. In India, it is cultivated at an altitude of 1,500 m in north-western regions that extend to hilly areas of Punjab, Himachal Pradesh, and Jammu.

3.10.3.1 Morphology

The Ashwagandha plant is an erect, enduring, more branched, woody shrub with a height of 1–3 m. Stems are branched, tubular, tough, stellate-hoary momentum. Roots are erect and unbranched, and the thickness of root depends on age; the secondary roots are in the form of fibres, which are externally buff to grey yellow in colour with longitudinal crinkles. Crown has two to six thickened stem base and nodes are found on the places where petiole ascends. Petioles are green, cylindrical, fractured, short, and uneven. Dry roots are cylindrical, tapering at apex with a brownish white surface outside and pure white inside when broken. Leaves are simple, exstipulate, 2–11 cm long and 1.5–9.0 cm wide, cauline and ramal, petiolate, lanceolate, apex acute or rounded, decurrent on vegetative shoots 8–10 cm long, reproductive shoots are 3–8 cm long and opposite. Leaves are pubescent on down surface unicostate, reticulate venation, arranged in pairs of one large and one small and slightly laterally. Leaf margin is wavy or entire. Inflorescences are axillary, dense, multi-flowered inflorescence with 2–25 yellow green short pedicellate flowers. Flowers are radially symmetrical, bell-shaped (campanulate), acute triangular lobes, the length of corolla is just double than calyx, 7–8 mm long with lanceolate lobes, 5 stamens, slightly exerted, petals are moderately fused with corolla. Ovary was superior, glabrous, stigma is slightly bifurcated. Fruits are spherical 5–6 mm in diameter, orange red in colour, surrounded by green membranous, inflated calyx about 2.5 cm in diameter and slightly 5 angled. Seeds are 2.5 mm in diameter and have light yellow colour. The roots have the characteristic odour of horse sweat, taste is sweetish, bitter, astringent, and slightly mucilaginous.

3.10.3.2 Microscopic

The TS of root appears a slender band of yellowish cork, thin cortex packed with starch grains, exfoliated or crushed, cork cambium has 2–4 diffused cells layers, and secondary cortex is made up of 24–26 layers of dense parenchyma cells. In vascular bundle, phloem is parenchymatous, consisting of sieve tube, companion cells. Xylem and phloem get separated by four to five cell-layered elongated cambium rings. The xylem is hard and the whole vascular bundle is separated by multiseriate medullary rays.

3.10.4 Phytochemistry

Withania comprises steroidal lactones, namely "withanolides," including withaferin A, 27-deoxywithaferin A, withanolides I–XI, and withasomniferols A–C. It also consists of alkaloids, including anaferine, anahygrine, cuscohygrine, DL-isopelletierine, 3-tropyltigloate, and tropine. Other compounds are sitoindosides VII–X, withaferin A, and acyl steryl glucosides served as antistress agents. The aerial parts of *W. somnifera* consists of 5-dehydroxy withanolide-R and withasomniferin-A. Ashwagandha is also rich in microelements, mainly iron. Therapeutic active compounds sitosterol, daucosterol, withaferin-A, 2,3-dihydrowithaferin A-3β-*O*-sulphate, and withasomniferol-A mainly exhibit immune modulatory activity.

3.10.5 Pharmacological Properties

The Indian ginseng rejuvenates the body cells that slowed the ageing process, boosts the body's defence mechanism, and revitalizes the body in stressful conditions. It can be utilized in all age groups and for both genders, even during pregnancy, without any side effects.

3.10.6 Scientific Investigation for the Management of Cancer

Numerous studies have reported that *W. somnifera* has anti-cancer properties in a variety of cancer types, including leukaemia, melanoma, prostate cancer, breast cancer, ovarian cancer, head and neck cancer, and colon cancer. It is a plant that has been used extensively in Ayurvedic, Siddha, and Unani medical systems since ancient times. It has recently been shown in experimental models to have anti-tumourigenic effects. The plant's roots have been shown to have anti-tumour and anti-preventive properties. Various parts of *W. somnifera*, particularly the roots, have been shown to be effective against various types of cancer. The most active components with anti-cancer activity include withanolides and withaferins, as well as a few other metabolites such as withanone and withanosides, which have been shown to be effective against various cancer cell lines.

3.10.7 Reported Mechanism of Action as Anti-cancer Activity

The precise processes by which *W. somnifera* and its withanolides induce apoptosis are unknown. However, data from several publications suggest that the drug's anti-cancer activity may be demonstrated by increased expression of pro-apoptotic genes as well as suppression of proliferative pathways. In order to establish its anti-cancer activity, withaferin A inhibits cell proliferation, invasion, metastasis, angiogenesis, proteasome, endoplasmic reticulum stress, protein folding, and maturation in cancer cells and regulates multiple targets through direct interaction or regulation of secondary targets.

3.10.8 Toxicological Aspects

W. somnifera has been regarded a safe medicine for the treatment of cancer, with the acute LD_{50} value in rats being 465 mg/kg (332–651 mg/kg) and in mice being 432 mg/kg (229–626 mg/kg). These tests proved that this plant is non-toxic in a wide range of tolerable levels, and it may be considered that the doses recommended for its formulations in humans are safe.

3.10.9 Conclusion and Future Aspects

W. somnifera possesses antistress, anti-inflammatory, anti-cancer, and immunostimulatory properties. Its treatment and its effect on tumour growth inhibition in TRAMP and Pten-knockout prostate cancer mouse models with intact immune systems demonstrate increased anti-tumour immunity. The *W. somnifera* induces Th1-type responses and NK cell activation, the slow-growing nature of some prostate cancers may provide an opportunity for immune manipulation at early stages of prostate cancer and may represent a paradigm for the use of *W. somnifera* for cancer-related or chemotherapy-induced fatigue, cancer prevention, and therapeutic efficacy. WFA use in conjunction with traditional treatment regimens may aid in the induction of both innate and adaptive immune system arms, resulting in improved therapeutic efficacy against prostate cancer.

3.11 *MENTHA SPICATA*

3.11.1 Background

M. spicata L. is also known as spearmint and menthe (mint). It belongs to the Lamiaceae family. It has 38 species throughout the world. Most of *Mentha* species contain essential oils. It is very useful as a flavouring agent; it has fragrance of many herbal teas and is also used for the enhancement of taste and flavours of many dishes. It is also used for relaxing the mind for some patients. Some *Mentha* species like *M. condensis* L., *M. aquatica* L., *M. spicata* L., and *M.* × *piperita* are the commercial valuable species which are utilized for the production of the essential oil.

Kingdom	**Plantae**
Phylum	Magnoliophyta
Class	Magnoliopsida
Order	Lamiales
Family	Lamiaceae
Genus	*Mentha*
Species	*spicata*
Binomial name	*M. spicata* L.

3.11.2 Ethnomedical Considerations

M. spicata is very popular in different tribal areas as folk medicine for the management of gastrointestinal problems like flatulence, nausea, anorexia, and liver-related problems (Figure 3.11). Leaves of *M. spicata* have essential oil which has therapeutic potential and is used as carminative tonic, stomach tonic, and as anti-cough, astringent, analgesic, and sedative. In the traditional system of medicine, it is used in the treatment of diarrhoea, bloating, abdominal pain, indigestion, intestinal weakness, cough and cold, sinusitis, influenza headache, nose bleeding, and psychological problems in children.

3.11.3 Morphology and Pharmacognostic Characteristics

M. spicata L. is also known as spearmint, brown mint, garden mint, lady's mint, sage of Bethlehem, and menthe (mint). It belongs to the Lamiaceae family. It has 38 species throughout the world. It is indigenous to Northern England. It is distributed in tropical regions of the world like Europe, North America, India, China, South Africa, and Brazil.

3.11.3.1 Morphology

M. spicata L.is creeping perennial rhizomatous herb, glabrous, and strong aromatic odour shrub. Its height is up to 30–100 cm, has hairless stem and foliage, and it has fleshy underground rhizome. Leaves of *M. spicata* L. are ovate to lancolate (5–9 cm in length and 3 cm in width) and have serrated margin. Its flowers are pink or white in colour and have slender spikes (2.5 mm in length and 3 mm in width). Its stem is square-shaped and it is a special characteristic of mint species.

FIGURE 3.11 Leaves of *Mentha spicata* L.

3.11.3.2 Microscopic Characteristics

M. spicata has simple leaves and pinnate venation. *M. spicata* lamina exhibited different size of polygonal epidermal cells (upper) and has epidermal hairs and simple hair-glandular trichomes (peltate scale). Mesophyll cells of the leaf of *M. spicata* are dorsoventral which is formed of palisade layers and parenchyma (spongy). Below the upper epidermis is palisade layer and above the lower epidermal the spongy parenchyma cells are present. It contains collateral vascular bundles; lower region has phloem, formed of sieve tube cells, companion cells, and phloem parenchyma. Xylem is formed of radial vessels. The metaxylem is present in the lower region and protoxylem is present in the upper region which is separated by xylem parenchyma.

3.11.4 Phytochemistry

M. spicata L. contains main active phytoconstituents and essential oils like carvone (49.62–76.65%) (it is responsible for the smell of *M. spicata*), limonene (9.57–22.3%), and 1,8-cineole (1.32–2.62%).Yield of the essential oil is varied, between 2.41% and 2.74%. The other main active phytoconstituents are also found in the essential oil of *M. spicata* like pulegone (26.7–29.6%), piperitone (22.2–28.2%), limonene (3.2–5.2%), α-phellandrene (1.3–2.6%), *trans*-caryophyllene (5.2–8.0%), and germacrene D (3.08–5.32%).

3.11.5 Pharmacological Effects

M. spicata extracts and essential oils were investigated for its various pharmacological properties which show antioxidant, anti-cancer, antiparasitic, antimicrobial, and antidiabetic effects.

3.11.6 Scientific Investigation for the Management of Cancer

The Lamiaceae and Asteraceae families are among the medicinal plants that have many biological effects. *M. spicata* is a herb with several biological properties. Cytotoxicity of essential oils of *M. spicata* on some cancer cells has been reported. *M. piperita*, *M. spicata*, *M. aquatica*, *M. crispa*, *M. pulegium*, and *M. longifolia* have been checked for their cytotoxic properties. The results showed that the herb exhibits concentration-dependent effect against the cancer cell lines. Studies also reported antiproliferative effects of *M. spicata* on particular breast cancer cell lines, fibrosarcoma, leukaemic monocyte, and mouth epidermal carcinoma.

3.11.7 Reported Mechanism of Action

Many studies have evaluated its antioxidant activity either by measuring its effectiveness in scavenging free radicals or by directly assaying the products formed using photometric techniques. The oil exhibits potent radical scavenging activities.

3.11.8 Toxicological Aspects

Toxicological investigations of *M. spicata* show the safety of this species at different doses and several periods of use which justify its use in traditional medicines. From the toxicological studies performed on the drug, it can be declared that *M. spicata* is an experimentally safe plant. However, prolonged treatment in high doses can lead to specific problems.

3.11.9 Conclusion and Future Aspects

M. spicata may have anti-tumour properties. *In vivo* research, as well as the discovery of potent *M. spicata* components with anti-cancer activity and their precise mechanism of action, could aid in the development of new anti-cancer therapeutic medicines.

3.12 *ABELMOSCHUS ESCULENTUS* L.

3.12.1 Background

A. esculents L. is also known as okra and ladies' fingers and it belongs to the Malvaceae family. It is an adorable and important medicinal plant. It is used as a vegetable crop only. It is mainly found in the Indo-Pak subcontinent. It is located mainly in the tropical and warmer regions of Asia. It contains a lot of medicinal classes of active phytoconstituents such as polyphenols, flavonoids, mucilaginous, tannins, sterols, triterpenes, mucilage, vitamin C and micronutrients (Ca, P, Fe), beta carotene, and vitamin B. Its fruits, leaves, and seeds are used for the health benefits.

Its fruits are mainly consumed as vegetables. Its seed is used as a rich source of protein, supplement, and fortified food.

Kingdom	**Plantae**
Phylum	Magnoliophyta
Class	Magnoliopsida
Order	Malvales
Family	Malvaceae
Genus	Abelmoschus
Species	*A esculents* L.
Binomial name	*Abelmoschus esculents* L.

3.12.2 Ethnomedical Considerations

A. esculentus L. is a medicinal and a very nutrient-rich plant (Figure 3.12). Its roots are a major source of mucilage, so it shows demulcent action. Its roots are used for the management of the syphilis. The root juice is utilized for the management of wound-healing, boils, and cuts. Decoction of immature capsules is used as diuretic, demulcent, and emollient. It is also used in the management of dysuria and gonorrhoea. The leaves of *A. esculentus* L. are used to furnish an emollient poultice. The seeds of *A. esculentus* L. are used as antispasmodic, cordial stimulants. The roasted

FIGURE 3.12 Whole plant of *Abelmoschus esculentus* L.

seed infusion is used for its sudorific properties. Commercially, it is distributed in Yugoslavia, Pakistan, West Bengal, Burma, Japan, India, Turkey, Iran, West Africa, Malaysia, Brazil, Ghana, Bangladesh, Afghanistan, Ethiopia, Cyprus, and the southern United States.

3.12.3 Morphology and Pharmacognostic Characteristics

3.12.3.1 Morphology

A. esculentus L. is also known as okra, ladies' fingers, bhindi, and bamia and it belongs to the Malvaceae family. It is an adorable and important medicinal plant. It is used as a vegetable crop only. It is mainly found in the Indo-Pak subcontinent. It is found in the tropical and warmer regions of Asia. It is commercially distributed in India, Turkey, Iran, West Africa, Yugoslavia, Bangladesh, Afghanistan, Pakistan, West Bengal, Burma, Japan, Malaysia, Brazil, Ghana, Ethiopia, Cyprus, and the southern United States.

3.12.3.2 Macroscopic Characteristics

A. esculentus L. is a herbaceous hairy annual plant. Leaves of *A. esculentus* L. are heart shaped. It is found in three to five lobes. Its flower is yellow and crimson in the centre. The fruits can be purple, dark green in colour and it has a hairy base. A tapering 10-angle capsule is 10–25 cm in length. It contains a lot of oval dark-coloured seeds.

3.12.4 Phytochemicals and Constituents

It contains a lot of medicinal classes of active phytoconstituents such as multicellular fibres, alpha-cellulose (67.5%), hemicellulose (15.4%), lignin (7.1%), pectin matter (3.4%), fatty acid (3.9%), waxy matter, polyphenols, flavonoids (quercetin), musilagenous, tannins, sterols, triterpenes, mucilage, vitamin C, micronutrients (Ca, P, Fe), beta carotene, and vitamin B.

3.12.5 Pharmacological Properties

A. esculentus contains various nutrients and important phytochemicals. It shows different biological activities like antibacterial, antioxidant, anti-inflammatory and immunomodulatory, antidiabetic, organ protective, and neuropharmacological activities.

3.12.6 Scientific Investigation for the Management of Cancer

Polysaccharides, polyphenols, and flavonoids have previously been found to be abundant in okra pods. The latter two have substantial antioxidant effects and are derived from okra seeds, whereas its skin extract did not exhibit such reactions. Isoquercitrin, a substance found in okra seed extract, has a higher bioavailability than quercetin and has been shown to have a number of chemoprotective effects *in vitro* and *in vivo* against oxidative stress, cardiovascular disorders, diabetes, allergic reactions, and cancer. Isoquercitrin has been demonstrated to slow the progression of urinary bladder and pancreatic cancers, as well as colon cancer suppression.

3.12.7 Reported Mechanism of Action as Anti-cancer Activity

The antioxidant activity of leaf ethyl acetate extract is comparable to that of ascorbic acid. The presence of glycosidic flavonoids, including quercetin 3-sambubioside and isoquercitrin, has been linked to the plant's antioxidant potential. The existence of flavonoid chemicals in okra has been proven, and direct delivery has the greatest lethal effect on breast cancer cell lines, followed by hepatocellular carcinoma and cervical carcinoma cell lines. Cytotoxic effects on cancer cell lines show a dose and time-dependent suppression of cell proliferation and migration, possibly due to VEGF production inhibition, leading to apoptosis and cell death, possibly due to flavonoid chemicals found in the plant. Okra is used as a chemopreventive drug because it inhibits cancer cell proliferation through correct signalling processes. Metabolic products derived from the seeds and flowers of *A. esculentus* have been studied in the treatment of many cancers.

3.12.8 Toxicological Aspects

Toxicological studies on *A. esculentus* reveal that this species is safe at various doses and for extended periods of time, justifying its usage in medicines. According to the toxicological research conducted on the medicine, the plant is experimentally safe.

3.12.9 Conclusion and Future Aspects

Consumption of okra fruit and/or leaf extracts may add significantly to antioxidant activity. It is possible to conclude that this compound has the potential to serve as a source of useful anti-cancer drugs and to be used to improve health.

3.13 *PHYLLANTHUS AMARUS*

3.13.1 Background

P. amarus Schum. & Thonn. is commonly known as Bhumyaamalaki. It belongs to the Euphorbiaceae family. It is distributed throughout the tropical and subtropical regions of the world. In the traditional system of medicine, it is used for kidney problems, diabetes, pain, jaundice, skin ulcer, gonorrhoea, chronic dysentery, and hepatitis B. *P. amarus* contains different classes of active phytoconstituents like lignans, alkaloids, phenolic, tannins, flavonoids, sterols, phyllanthin, hypophyllanthin, corilagin, geraniin, and volatile oils. *P. amarus* has a wide area of uses such as immunomodulatory, antiviral, antibacterial, antioxidant, antiulcer, anti-cancer, anti-inflammatory, antispasmodic, antinociceptive, anti-hyperglycaemic, antilipidemic, aphrodisiac, contraceptive, and antiamnesic activity.

Kingdom	Plantae
Phylum	Magnoliophyta
Class	Magnoliopsida
Order	Euphorbiales
Family	Euphorbiaceae
Genus	*Phyllanthus*
Species	*amarus*
Binomial name	*Phyllanthus amarus* L.

3.13.2 Ethnomedical Considerations

The genus *Phyllanthus* has more than 1,000 species (Figure 3.13). All these species contain a lot of active phytoconstituents which are responsible for therapeutic activity. *P. amarus* is distributed across India: Maharashtra, Uttar Pradesh, Punjab, Bihar, Orissa, Andhra Pradesh, some parts of Madhya Pradesh, Karnataka, and Bengal.

3.13.3 Morphology and Pharmacognostic Characteristics

3.13.3.1 Morphology Characteristics

P. amarus is a small tropical annual herb (60–75 cm tall) and glabrous. It is grown in moist soil and requires light shade and its complete maturity comes in two to three months.

Leaves: Sub-sessile, numerous, stipulate, oblong, small leaflets (paripinnate) and base is round.

Stem: Branched and angular in shape.

FIGURE 3.13 Whole plant of *Phyllanthus amarus*.

Flower: It is star-shaped and around 2 mm in size.
Root: Its root is firm, twisted, and woody.

3.13.3.2 Microscopic Characteristics

The observations of the studied microscopic characteristics are compiled in a comparative manner with respect to three important parts of the plant: root, stem, and leaf.

a. *Root*
    - The TS of root shows epidermis as a single layer of thin-walled cells.
    - The cortex region has six to eight layers of parenchymatous cells without intercellular spaces.
    - The inner cortex consists of patches of macrosclereids.
    - The vascular cylinder consists of 5–8 layers of secondary phloem cambium and 25–40 layers of secondary xylem along with fibres (pits rare, bordered; ends tapering; wall tetra-to hexagonal) and vessel members (long with tails at both ends, pits circular, bordered; perforation plate simple).
    - The xylem parenchyma is thin-walled with uniseriate rays, three to eight cells high, usually heterogeneous type, while the pith is parenchymatous.

b. *Stem*
    - The TS of the stem is circular in outline and shows central pith occupying the major area of the section, encircled by continuous band of xylem and a ring of discontinuous periclyclic fibres, narrow parenchymatous cortex, and a layer of epidermis and collenchymatous narrow hypodermis.
    - Phloem is narrow, parenchymatous, cambium is distinct, xylem consists of radial rows of vessels tracheids, thin-walled fibres, parenchyma, and uniserate to biseriate medullary rays; pith is wide and parenchymatous; cells get disintegrated on drying, developing cavity in the centre, cluster and rosette crystals of calcium oxalate throughout the parenchymatous cells of the cortex and the pith.

c. *Leaf*
    - The TS of the leaf passing through midrib is slightly elevated on the lower side and flat on the upper side.
    - It shows a layer of upper epidermis, its cells being bigger in size than the lower one and covered with thin cuticle.
    - The important cellular characteristics of the leaf are type of stomata cell wall, margin of lamina, nature of crystals, and nature palisade tissue in midrib.

3.13.4 Phytoconstituents

P. amarus contains different classes of active phytoconstituents like lignans (phyllanthine and hypophyllanthine, alkaloids (ent-norsecurinine, sobubbialine, epibubbialine; diarylbutane, nyrphyllin and a neolignan, phyllnirurin), phenolic tannins (hydrolysable tannin: phyllanthusiin D amariin amarulone, amarinic acid), flavonoids (quercertin, astralgin, quercertrin, isoquercitrin and rutin), sterols, phyllanthin, hypophyllanthin, corilagin, geraniin, and volatile oils.

3.13.5 Pharmacological Properties

P. amarus is used as antiseptic, antidiabetic, antihypertensive, analgesic, anti-inflammatory, antimicrobial, and anti-tumour activity. Decoction of *P. amarus* is utilized as herbal bath and in crams, asthma, uterus problems, and stomach ache. It is also used as an appetizer and as a tonic. Decoction of leaves of *P. amarus* is used as diuretic and in diabetes, dysentery, hepatitis, skin disorder and menstrual disorder. *P. amarus* extracts are utilized as blood purifiers and in anaemia and malaria.

3.13.6 Scientific Investigation for the Management of Cancer

P. amarus aqueous extract exhibited potent anticarcinogenic activity against 20-methylcholanthrene-induced sarcoma. *P. amarus* has anti-proliferative activity against various cancers, implying that it may be able to regulate cell proliferation.

3.13.7 Reported Mechanism of Action as Anti-cancer Activity

P. amarus' anti-tumour and anti-cancer efficacy may be demonstrated by inhibiting carcinogen metabolic activation as well as cell cycle regulators. *P. amarus* has been shown to prevent tumour production by inhibiting *N*-methyl *N*-nitrosoguanidine (MNNG)–induced gastric carcinogenesis.

3.13.8 Toxicological Aspects

Toxicological studies on *P. amarus* revealed that the herb had potentially harmful effects on the blood, so caution should be exercised when using *P. amarus* as a medicinal plant.

3.13.9 Conclusion and Future Aspects

The ability of *P. amarus* to alleviate pain is dose and duration dependent. The extracts reduce mutagenicity and genotoxicity significantly. However, some negative effects have been observed in toxicological studies. As a result, the herb can be used to treat cancer, but only with caution.

3.14 *MALUS DOMESTICA*

3.14.1 Background

M. domestica, the common name apple, is obtained from *Malus pumila* Mill. belonging to the family Rosaceae. It is one of the most economically and traditionally significant, nutrient-rich fruits grown in all temperate zones. The whole fruit is edible excluding seeds; apart from that, many other products are formed from them: ciders and juices, jams, compotes, tea, wine, or dry apples. The apple tree is fragile and has simple flower clusters. The fruit is called "pome." Apple trees are grown-up in all temperate climates around the world, and the fruit is widely available in commercial markets. About 2,500 known varieties of apples (plants) are grown in the United States and more than 7,500 varieties are grown worldwide.

The cultivated apple is thought to have originated in central Asia from the wild species *Malus sieversii* (Ledeb.) M. Roem.

Kingdom	**Plantae**
Phylum	Magnoliophyta
Class	Magnoliopsida
Order	Rosales
Family	Rosaceae
Genus	*Malus*
Species	*domestica*
Binomial name	*Malus domestica* L.

3.14.2 Ethnomedical Considerations

An apple tree is an abscission tree, usually standing 6–15 ft (1.8–4.6 m) tall in planting and up to 30 ft (9.1 m) in the wild. The leaves are alternately arranged in light green ovals with circular edges and slightly lower (Figure 3.14).

3.14.3 Morphology and Pharmacognostic Characteristics

3.14.3.1 Morphological Characteristics

1. Blossoms are produced in the spring along with the leaves budding on spurs and long shoots.
2. The flowers are pinkish white and have five petals. An inflorescence consists of a cyme with four to six flowers.

FIGURE 3.14 Photograph of *Malus domestica*.

3. The flowers in the middle of the inflorescence are called the "king bloom"; it opens first and can develop a superior fruit.
4. The fruit ripens in late summer or autumn, and cultivars come in a variety of sizes.
5. The skin of ripe apples is usually red, pink, yellow, green, or russet, although there are many varieties with two or three colours.
6. The skin of the fruit may also be completely or partially russet, which is rough and brownish.
7. The skin is enclosed in a shielding layer of epicuticular wax. Exocarp (flesh) is usually yellowish-white, although Exo carp in pink or yellow occur.

3.14.3.2 Microscopic Characteristics

The *M. domestica* had a dry and rough, greenish-yellow or yellow peel, covered with intense red, flaky-striped blush, and light grey lenticels. The surface of the fruit had a smooth, sticky, creamy yellow surface, sometimes covered with a weak, fuzzy-striped carmine blush and greyish lenticels. The epidermal cells of the *M. domestica* cultivar retain the meristematic properties and contribute to the expansion of the tissue and by dividing and extending along the periclinal walls, they contribute to expansion of the epidermal surface as the fruit volume increases by dividing and expanding along the perimeter wall. The trichomes that cover the surface of apples during fruiting form a mechanical defence and protect against abiotic and biotic agents. After their modification, this function is performed by cuticular and epicuticular waxes.

3.14.4 Phytochemistry

Apples contain high levels of polyphenols (up to 2 g/kg fresh weight) and other phytochemicals, many of which are strong antioxidants commonly used in the United States. Apples had the second highest level of antioxidants in the second place in phenolic compounds and had the highest amount of free phenolic.

3.14.5 Pharmacological Effects

- Traditional uses of apples include treatment for cancer, heart ailments, diabetes, fever, dysentery, constipation, scurvy, and warts.
- Preliminary evidence that the antioxidant and anti-inflammatory effects of apples may provide significance in a range of other situations.
- It may prevent cardiovascular disease by virtue of its beneficial effects on cardiovascular risk factors (e.g. atherosclerosis, obesity, diabetes); and may have beneficial effects on pulmonary function, including preventing asthma.

3.14.6 Scientific Investigation for the Management of Cancer

The apple (*M. domestica*), a member of the Rosaceae family, is a significant and widely grown and consumed fruit throughout the world. Apples, particularly their peels, have been found to have high antioxidant activity and can significantly inhibit the growth of liver and colon cancer cells. Several studies have shown that apples have potent antiproliferative activity. When colon cancer cells are treated with apple extracts, cell proliferation is inhibited in a dose-dependent manner.

3.14.7 Reported Mechanism of Action as Anti-cancer Activity

Apples possess a high concentration of phenolic compounds with antioxidant activity, and there is a link between the number of phenolic compounds and antioxidant activity across all apple cultivars. Catechin, chlorogenic acid, epicatechin, cyanidin-3-galactoside, procyanidin, coumaric acid, gallic acid, phloridzin, quercetin-3 galactoside, and quercetin-3-rhamnoside are only a few of the antioxidant chemicals found in apples and apple peels. However, several chemicals, such as epicatechin, catechin, phloridzin, and procyanidin, are found in lower concentrations in apple flesh than in peels.

3.14.8 Toxicological Aspects

Toxicological tests on *P. amarus* indicated that the herb has potentially harmful effects on the blood; hence caution should be given when using *P. amarus* as a medicinal plant.

3.14.9 Conclusion and Future Aspects

The ability of *P. amarus* to alleviate pain is dose and duration dependent. The extracts diminish mutagenicity and genotoxicity considerably. However, certain negative effects have been reported in toxicological tests. As a result, the plant can be used to treat cancer, but only with caution.

3.15 *SILYBUM MARIANUM* L.

3.15.1 Background

Silybum is called a tree plant derived from the Greek silybon or silybos, meaning "twig" or "stem." Milk seeds (*S. marianum* L. Geert., Asteraceae) have been used for centuries as herbal medicine, especially to treat liver disease.

Milk thistle is inherent to the Mediterranean region and it has also spread to East Asia, Australia, and America.

Kingdom	**Plantae**
Division	Magnoliophyta
Class	Magnoliopsida
Order	Asterales
Family	Asteraceae
Genus	*Silybum*
Species	*marianum*
Binomial name	*Silybum marianum* L.

3.15.2 Ethnomedical Considerations

Milk thistle (*S. marianum*) is a herbal product commonly used by HIV-positive patients to control or prevent hepatotoxicity caused by antiretrovirals or hepatitis B or C (Figure 3.15). It is used to treat liver problems, such as hepatitis, and cancer of the bladder. It contains a variety of lignans, including silandrin, silybin, silychristin, silydianin, silymarin, and silymonin. Anaphylactic shock has been reported after the use of a herbal tea containing an extract of the fruit of the milk thistle. It inhibits CYP3A4 and uridine diphospho-glucuronosyltransferase in human hepatocyte cultures.

3.15.3 Morphology and Pharmacognostic Characteristics

3.15.3.1 Morphological Characteristics

Milk thistle is a tall plant with milky white veins, dark green leaves, stem thorns, and purple flowers. Milk thistle pollen grain shape is prolate when looking at the equatorial region and seems semi-angular in polar sight. The seeds have white pappus brownish in colour, shiny, taking hard skin achenes, mainly 6–8 mm long.

FIGURE 3.15 Photographs of *Silybum marianum*. (A) Flower of *Silybum marianum*, (B) Leaves of *Silybum marianum*, and (C) Seed powder of *Silybum marianum*.

3.15.3.2 Microscopic Description

3.15.3.2.1 Stems

Milk thistle has a stem 40–200 cm in height, glabrous or slightly downy, erect, and branched in the upper part of the plant.

3.15.3.2.2 Leaves

The basal leaves are alternate, large, and glabrous with spiny margins. The leaves can be 50–60 cm long and 20–30 cm wide.

3.15.3.2.3 Inflorescences

The inflorescences are enclosed by spiny bracts. The florets are hermaphrodite.

3.15.4 Phytochemistry

Silymarin is a lipophilic extract from the milk thistle and is composed of isomer silybin, silydianin, flavonolignans, and silychristin. Milk thistle seeds contain small amounts of taxifolin and about 20–35% fatty acids as well as polyphenolic compounds.

Silymarin represents 1.5–3% of the weight of dried fruit and is an isomer compound of unique flavonoid complexes—flavonolignans. Silymarin is silybin, isosilybin, silychristin, iso-silychristin, silydianin, and silimonin. The chemical composition of milk thistle fruit is flavonolignans and it combines with other flavonoids (such as taxifolin, quercetin, dihydrokaempferol, kaempferol, apigenin, naringin, eriodictyol, and chrysoberyl), 5,7-dihydroxy chromone, dihydro-coniferyl alcohol (60%) acid (30%, oleic acid, 9% palmitic acid), tocopherol, sterols (cholesterol, campesterol, stigmasterol, and sitosterol), sugar (arabinose, rhamnose and xylose), and protein.

3.15.5 Pharmacological Properties

Milk thistle was used for treatment. Its flowers, leaves, and roots have been used in European diets as vegetables, and its achene is used as a coffee. Silymarin and silybin are helpful for fighting against hepatotoxins such as alcohol, tetrachloromethane, toluene, xylene, and carbon tetrachloride. Silybin acted as a stimulating agent for DNA polymerase reactions and cell regeneration of liver and ribosomal RNA. Silybin regulated the production of free radicals, increasing glutathione peroxidase and superoxide dismutase, stabilizing membranes and increasing the glutathione in liver and human cell lines. Liver inflammation is controlled by repressing the pathway of 5-lipoxygenase.

3.15.6 Scientific Investigation for the Management of Cancer

Milk thistle (*S. marianum*) is a well-known herb used to treat liver and biliary problems. *S. marianum* was also used as cancer therapy for prostate, skin, breast, cervix, and hepatocellular carcinoma; however, different studies produced contradictory and inconsistent results. It has been proposed that silibinin could be utilized to treat non-melanoma skin cancer in people. Silibinin has been used to treat a variety of liver diseases and has been shown to have a significant inhibitory effect on a variety of tumours, including breast, oral, colon, prostate, and lung cancers.

3.15.7 Reported Mechanism of Action as Anti-cancer Activity

Several *in vivo* investigations have demonstrated that silymarin has antioxidant and anti-inflammatory properties that prevent skin cancer in an *in vivo* animal model. Silibinin, a key component of silymarin, has been demonstrated to suppress the synthesis of UVB-induced thymine dimers, accelerate DNA repair, and/or cause death in injured cells. Silymarin inhibits growth factors and weakens proliferative mediators in cancer cells by stimulating apoptosis and increasing tumour suppressor and cell cycle inhibitors.

3.15.8 Toxicological Aspects

Silymarin is the active ingredient that protects against drug toxicity, including chemotherapy, and is also a powerful antioxidant.

3.15.9 Conclusion and Future Aspects

Using naturally occurring medicinal herbs as an anti-cancer agent appears to be a common practice around the world. For example, silymarin has been shown in animal tumour models to have a protective effect against numerous types of carcinogenic agents.

3.16 *CITRULLUS COLOCYNTHIS* L.

3.16.1 Background

C. colocynthis L. Schrad. belongs to the Cucurbitaceae or squash family; the common name is bitter-apple and is cultivated in Europe for ornamental purposes (Figure 3.16). Vernacular names are colocynth, bitter gourd, bitter apple, bitter cucumber (English), and coloquinte (French). The principal therapeutic part of the plant is the fruit pulp.

Kingdom	**Plantae**
Division	Tracheophyta
Class	Magnoliopsida
Order	Cucurbitales
Family	Cucurbitaceae
Genus	*Citrullus*
Species	*colocynthis*
Binomial name	*Citrullus colocynthis* L.

FIGURE 3.16 Photograph of *Citrullus colocynthis* L.

3.16.2 Ethnomedical Considerations

It originated in the arid regions of North Africa and was known throughout the Mediterranean region since Bible times. This plant is now found in North Africa: Algeria, Egypt; Libya, Morocco, Tunisia; Northeastern Tropical Africa: Chad, Ethiopia, Somalia; East Tropical Africa: Kenya; West Tropical Africa: Mali; Asia: Kuwait, Saudi Arabia, Iraq, Jordan, Lebanon, Syria, Yemen, Afghanistan, Iran, Turkey, India, Pakistan, Spain; and Australia.

3.16.3 Morphology and Pharmacognostic Characteristics

3.16.3.1 Morphological Characteristics

Colocynth is a perennial herbaceous vine, with angular and rough stems. Stems are 0.5–1.5 m, procumbent, branched, angular, and hirsute. The root is fleshy. Leaves are rough, three to seven lobed, 5–10 cm long, the middle lobe sometimes ovate, sinuses open, flowers monoecious, solitary, peduncled, axillary, corollas 5-lobed; ovary villous. Fruit is nearly globular, 4–10 cm in diameter with somewhat elliptical fissures, about the size of a small orange, variegated green and yellow, becoming yellow when ripe, with a hard rind, the pulp light in weight, spongy, easily broken, light yellowish-orange to pale yellow, and intensely bitter. Seeds are numerous, ovoid, compressed, smooth, dark brown to light yellowish-orange, borne on the parietal placenta.

3.16.3.2 Pharmacognostic Characteristics

Microscopic examination revealed the presence of epiblema, cortex, phloem, medullary rays, xylem parenchyma, xylem vessels (cavity), root bark, starch cells, and pigment cells. Microscopy showed the presence of tracheids (pitted), fibre, sclereids, epiblema cell, collenchyma cell, trachea, vessel, cork cell, fragment of the trachea, tracheids thickening, fragment of the vessel, fragment of fibre, and fragment of vascular bundle.

3.16.4 Phytochemistry

Phytochemical analysis of plant extracts revealed the presence of carbohydrates, protein, separated amino acids, tannins, saponins, phenolic, flavonoids, terpenoids, alkaloids, anthranol, steroids, cucurbitacin A, B, C, D, E (α-elaterin), J, and L, caffeic acid, and cardiac glycosides.

3.16.5 Pharmacological Properties

Root was used for breast inflammation and joint pain; externally, it was used for ophthalmia and uterine pain. The fruit and root were then rubbed with water and applied to the intestines and acne. The root paste is used for enlarged children's stomachs. The fruit was also used in ascites, biliousness, jaundice, cerebral congestion, colic, constipation dropsy, fever, worms, and sciatica. The root is used in abdominal enlargement, cough, asthma, inflammation of the breast, ulcers, urinary diseases, and rheumatism.

3.16.6 Scientific Investigation for the Management of Cancer

Breast cancer is the most common and often diagnosed malignancy, with the highest cancer-related death rate in women worldwide each year. Medicinal herbs are known for being a rich source of secondary metabolites, and their potential to treat cancer is receiving a lot of attention. Methanolic extract of *C. colocynthis* leaves and fractions has anti-cancer activity against breast cancer cells in a dose-dependent manner. Characterization of cancer stem cell markers treated with the herb reveals considerable downregulation, indicating that the extract is quite effective at lowering stemness.

3.16.7 Reported Mechanism of Action as Anti-cancer Activity

The main component analysis indicates a possible link between the high omega-6 fatty acid content of natural Handal seed oil and the anti-cancer effects reported against the cancer cell lines examined. Significant overexpression of cyclin-CDK inhibitors (p21 and p27) and cell cycle checkpoint regulators (HUS1, RAD1, ATM) followed by downregulation of downstream cell cycle progression genes verified the cell cycle arrest (cyclin A, cyclin E, CDK2).

3.16.8 Toxicological Aspects

Various studies show that cancer cell viability was significantly reduced after 48 hours of exposure relative to human fibroblasts when treated with *Citrullus* oil using the MTS assay at concentration points up to 50 mg/ml, indicating a reasonable anti-cancer activity and safety profile against human skin fibroblasts. However, health concerns such as colic, diarrhoea, vomiting, and liver impairment have been regularly observed with the usage of *C. colocynthis*.

3.16.9 Conclusion and Future Aspects

The anti-cancer potential of *C. colocynthis* leaves extract has been demonstrated by inhibiting the proliferation and stemness of breast cancer cells. Data from growth and proliferation experiments, as well as expression modulation of numerous cell cycle genes, reveal that C. colocynthis anti-cancer action is mediated by cell cycle arrest. However, more extensive research is needed to identify, isolate, and purify the most active component implicated in *C. colocynthis*' anti-cancer and cell cycle arrest capabilities.

3.17 *Cucumis sativus* L.

3.17.1 Background

The cucumber originated in India but is now grown on many continents. *C. sativus* known as cucumbers is a member of the Cucurbitaceous family, like melons, squash, and pumpkins. Cucumbers are alkaline, non-starchy vegetables. They are a cooling food, especially when used in vegetable juices. These vegetables are very high in water and very low in calories. It is an excellent source of potassium, vitamin C, and folic acid (only if not raw).

C. sativus is a creeping vulture that grows on trellises or other supporting structures, covering the base with small flexible ropes. The root system consists of a taproot, which reaches 1 m in height. Leaf rotation, Slight Crop Profile adjustment, and skin care construction test 72-lobed, with five angles; they have crenate edges. A flowering male stalk with distinctive fruit, such as a short-length stem, 5-lobed hairy leaves or angles, 1–2 male flowers together in cages, and female flowers alone, both very hairy. The cucumber fruit is ovoid or oblong, almost cylindrical, with side and tapered ends, and can be up to 60 cm (24 in.) long, 10 cm (3.9 in.) wide, and fleshy.

Kingdom	Plantae
Division	Tracheophyta
Class	Magnoliopsida
Order	Cucurbitales
Family	Cucurbitaceae
Genus	*Cucumis*
Species	*sativus*
Binomial name	*Cucumis sativus* L.

3.17.2 Ethnomedical Considerations

C. sativus is believed to have originated in the southern Himalayan region of southern Asia (Figure 3.17). *C. sativus* is one of the most widely grown vegetable crops in the world. It is found in the Indian subcontinent. Cucumber was taken east and planted in China 2,000 years ago. Until recently, cucumbers were thought to have spread to the west for the first time, familiar with the ancient Egyptians, Greeks, Romans, and Jews.

FIGURE 3.17 Photograph of *Cucumis sativus* L.

3.17.3 Morphology and Pharmacognostic Characteristics

3.17.3.1 Morphology

Cucumis is a genus of twins and tendrils of the Cucurbitaceae family. *C. sativus* known as "cucumber" is believed to have originated in Asia and exists as wild cucumbers in India and closely related to species found in the eastern Himalayas. The cucumber is an annual monoecious plant with trailing or climbing, four stems with five angles up to 5 m tall, and small branches with simple twigs up to 30 cm tall. This plant is covered with scaberulous hair and the root system is very large. The leaves are distinct, simple, and scaly to a petiole 5–20 cm long. Lamina is ovate triangular in shape, 7–20 cm × 7–15 cm, palmately 3–7 lobed, deeply attached to the base, top, toothed, and hispidulous on both sides. The flowers are sexual, common, pentamerous, approximately triangular sepals, 0.5–1 cm long, broad campanulate corolla, lobes up to 2 cm long, and yellow. Sturdy flowers come from a 3–7 fascicle with flowers with a pedicel length of 0.5–2 cm. Pistillate flowers are solitary, with short and thick pedicels up to 0.5 cm long, 5 cm long, ovary inferior, ellipsoid, muricate, 2–5 cm long, prickly hairy or warty, stigma 3-lobed. The fruit is almost cylindrical, long with sharp edges, and can be as large as 60 cm long and 10 cm wide. The fruit closes the seed and grows on the flowers.

3.17.3.2 Microscopic Characteristics

The cucurbit phloem is a complex with large sieve tubes on both sides of the xylem (collateral phloem) and extravascular elements that form a complex network connecting the rest of the vasculature. The vascular bundles of the stem interconnect at the node, facilitating lateral transport around the stem. The cucurbit phloem is complex, with large sieve tubes on both sides of the xylem.

3.17.4 Phytochemistry

The dietary value of cucumber is negligible, there being upwards of 96% water in its composition. The fruit's chemical composition at maturity is 4–5% solids, which include sugars, protein, fat (isokarounidiol, palmitoleic acid, heptadecanoic acid, karounidiol), phytosterols, phenolic acids, cellulose, potash, and cucurbitacins. The cucumber contains vitamins C, B1, and B2, provitamin A, organic acids, pectins, and essential oils. Glycosides, steroids, flavonoids, carbohydrates, terpenoids, and tannins were identified in an aqueous extract of the cucumber fruit.

3.17.5 Pharmacological Properties

In Ayurveda, many medicinal properties have been mentioned in cucumber. Various parts of plants such as seeds and leaves are used. The leaves are boiled and mixed with grated cumin seeds and powdered and used for throat infections. It is used as a refrigerator and is given for removing fever and inflammation. Cucumber is used for thirst, blood disorders, anuria, and inflammatory disorders. A cold drink prepared from seeds is used in the summer to have a cooling effect. Seed oil is used for fever, insomnia, and frontal headaches. Cucumber fruit is used as a demulcent, depurative, emollient, purgative, and solvent. Fresh fruit is used internally for the treatment of sensitive skin and fever and externally for burns, sores, and wounds; it is also used as a cosmetic to soften and whiten the skin. Raw cucumber fruit is used to treat celiac disease; in Indo-China, the unripe fruit is used to treat diarrhoea in children. The seeds are used as a diuretic, tonic, anthelmintic, and taeniacide. The leaf juice is emetic and is used to treat dyspepsia in children. A decoction of the roots is used as a diuretic.

Cucurbitin-1, a cell-permeable, bitter triterpenoid compound from *C. sativus* displayed anti-proliferative and anti-tumour properties in both *in vitro* and *in vivo* studies. In another study, cucurbitins also exhibited a wide-ranging *in vitro* and *in vivo* pharmacological effects like purgative, anti-inflammatory, and antifertility activities. Other studies showed that cucurbitin-1 may be potent chemoprotective agent for nasopharyngeal carcinoma with anti-invasion and anoikis-sensitizing activities.

3.17.6 Scientific Investigation for the Management of Cancer

C. sativus has anti-cancer properties. The methanolic extract was discovered to have potent cytotoxicity against cancerous cell lines, with MCF7 values. Aromatic plants contain a high concentration of anti-cancer compounds that are less harmful to normal cells. As a result, more emphasis has been placed on identifying innovative anti-cancer pharmaceutical medications.

3.17.7 Reported Mechanism of Action as Anti-cancer Activity

There are no such crucial modifications detected in *C. sativus* methanol extract (CSME) treated normal cells, indicating that the chemical is non-toxic to normal breast cell lines. The MTT assay of the chemical isolated from ethyl acetate fractions of *C. sativus* flowers demonstrates anti-cancer activity at all doses. Most anti-cancer substances or medications are designed to eliminate rapidly reproducing malignant cells, indicating cytotoxicity and inducing apoptosis in cancer cells.

3.17.8 Toxicological Aspects

According to a research, there are no such critical alterations seen in *C. sativus* methanol extract–treated normal cells, indicating that the compound is non-toxic to normal breast cell lines.

3.17.9 Conclusion and Future Aspects

More research is needed to determine the identity of the chemical component responsible for the anti-cancer activity which will help in the development of valuable anti-cancer drugs.

3.18 *ECBALLIUM ELATERIUM* L.

3.18.1 Background

The beautiful elaterium is brightly coloured, bright, flat, about 1/2 or 1 row, pale grey, green, or yellow, with a faint animal odour (slightly as tea-like Pharmacographia), and tastes very spicy and sour. The leaves are soft and snowy, although the roots are very hard and the plants can survive very cold winters. As the seed matures, this pressure forces the fruit to separate from the plant, expelling its seeds far away. The plant sometimes sow itself and can be a weed in warmer climates. It is under legal control as a weed in Australia.

Kingdom	**Plantae**
Division	Tracheophyta
Class	Magnoliopsida
Order	Cucurbitales
Family	Cucurbitaceae
Genus	*Ecballium*
Species	*elaterium*
Binomial name	*Ecballium elaterium* L.

3.18.2 Ethnomedical Considerations

The plant *E. elaterium* is commonly called squirting cucumber in English, *spritzgurke* in German, and *Concombre sauvage* in French and was known to the ancient Greeks, having been described by the Greek philosopher Theophrastus in his history of plants. It is a wild medicinal plant indigenous and found abundantly in the Mediterranean region, and cultivated in Central Europe and England. The fruit of this species is ovoid, fleshy; approximately 4 cm in length; when unripe, it is pale green in colour, and covered with numerous, uniseriate glandular hairs (Figure 3.18).

FIGURE 3.18 Photograph of *Ecballium elaterium* L.

3.18.3 Morphology and Pharmacognostic Characteristics

3.18.3.1 Morphology

This plant forms stems of thick flesh that lies on the ground and spread out on the exterior. These curved stems reach a height of 80 cm or more, but they do not rise and the plant stays low at about 25 cm. Stems branch only from the base section to the other smaller branches. The whole plant, except the flowers, is covered with a thick coat of white, strong, but non-spiny, bristles.

The leaves grow elsewhere near the stem and have a thick and long stem. Unlike other members of the gourd family, this plant has no tenders. The shape of the leaves is triangular and the base is cast like a heart (two circular lobes) or a lance (two pointed lobes). All the leaves are thick and fleshy and pale green on the upper side and white on the underside. The genes have waves, and they form small bumps or irregular teeth. The leaves grow up to 10 cm tall. New leaves or flowering stems grow only on the axils of mature leaves.

The flowering stems reach a height of 2–4 cm and are usually collected as two stems from the axil of a common leaf—one stem with a collection of four to six male flowers and one with a single female flower.

3.18.3.2 Microscopic Characteristics

The leaves of *E. elaterium* are homobaric, coarse, slightly curled in the margins, and reversely triangular in shape. In cross sections, they appear dorsiventral with a distinct upper layer of sized palisade cells and a subjacent second layer not always discernible. Palisade cells occupy about 50% of the mesophyll width. Spongy parenchyma is rather compact and confined to the lower part of the mesophyll. The leaves are amphistomatic. Stomatal density, on the adaxial side, is 220 mm approximately, whereas that of the abaxial side was impossible to estimate due to the dense indumentum. Guard cells are levelled with the other epidermal cells. Substomatal chambers are present on either side. The stomata of both surfaces are of the anomocytic type. Abaxial epidermal cells are larger in size, possess thicker cell walls, and, uncommonly for a xerophyte, do not seem to accumulate secondary metabolites within their vacuole. Adaxial epidermal cells are smaller, possess a thinner cell wall, and are free of accumulated compounds as well.

3.18.4 Phytochemistry

It produces a class of triterpenoids known as cucurbitacins, widely distributed in all cucurbit species. Cucurbitacins are a group of tetracyclic triterpenoids derived from the cucurbitane skeleton and found primarily in the Cucurbitaceae family.

3.18.5 Pharmacological Effects

Elaterium is an energetic hydragogue cathartic, which acts violently in small doses, causing widespread inflammation of the stomach and intestines, characterized by vomiting, cramping pains, and severe diarrhoea. It should be remembered that a small dose relieves stomach upset. Seed oil of *E. elaterium* inhibits the proliferation of cancer cell lines HT29 and HT1080. Seed oil from *E. elaterium* is found to be a rich source of polyunsaturated fatty acids, especially linoleic and punicic acid. High levels of polyunsaturated/saturated fatty acids can have beneficial effects on cancer prevention and treatment. Linoleic acid has been reported to play an important role in preventing mammary carcinogenesis. It also suppresses the proliferation of human breast adenocarcinoma (MDA-MB468) cells in SCID mice. In addition, it induces apoptosis in the lines of breast cancer cells. So *E. elatrium* can be considered a promising source of anti-cancer agents found in plants. β-Sitosterol, the major phytosterol in *E. elaterium* seed oil, shows activity to inhibit proliferation against colonies of cancer cells COLO 320 DM. In addition, *Û-tocopherol* shows great potential for inhibiting the growth of prostate and lung cancer cells by disrupting the formation of the sphingolipid system. This suggests that these natural compounds found as the major tocopherol forming *E. elaterium* seed oil may have the highest potential in preventing human cancer.

3.18.6 Scientific Investigation for the Management of Cancer

Cucurbitacins have a number of anti-cancer properties. Cucurbitacins D, E, and I purified from *E. elaterium* (L.) A. Rich. were studied. Cucurbitacins exhibit apoptotic and autophagy genes in the human gastric cancer cell line AGS (adenocarcinoma gastric cell line).

3.18.7 Reported Mechanism of Action as Anti-cancer Activity

Cucurbitacins D, E, and I isolated from the fruits of *E. elaterium* upregulate LC3 and induce sub-G1 cell cycle arrest and cell death in the human gastric cancer cell line AGS. Cucurbitacin I has a significantly greater effect on LC3 mRNA expression than cucurbitacins E and D.

3.18.8 Toxicological Aspects

Cucurbitacin B, derived from the *E. elaterium*, shows the anti-inflammatory activity as well as preventative and therapeutic properties against CCl4-induced liver damage. Cucurbitacin E extracted from *E. elaterium* was found to have immunomodulatory action. Cucumber squirting contains cucurbitacins, which cause diarrhoea and enteritis. It was thought to be abortifacient in the ancient world. *E. elaterium* may cause genotoxicity, allergic responses, renal and cardiovascular failure, keratoconjunctivitis, and well-defined oedematous regions in the eyes. The plant is considered highly toxic. It is not reasonable to use this plant internally, since its active properties differ a lot according to the season (it is very active in August). In addition, *elaterium* is very irritating for the digestive tract, so it should not be used by elderly people or those with intestinal or kidney problems. It is also abortive. It causes diarrhoea, haemorrhages, and kidney injuries. It is lethal in big doses. Extract of the plant, especially the juice found in the fruit, is rich in *elaterium* which is used as a drug that mainly has a powerful hydragogue cathartic action, and in large doses excites nausea and vomiting. If administered too frequently, it operates with great violence on both the stomach and bowels producing inflammation and possibly fatal results.

3.18.9 Conclusion and Future Aspects

Cucurbitacins are highly oxidized tetracyclic triterpenoids with cytotoxic properties against a variety of cancer cell types. Cucurbitacins are claimed to be JAK/STAT pathway inhibitors; however, their cytotoxic effects may involve other mechanisms such as the MAPK pathway, PARP cleavage, active caspase-3 expression, decreasing pSTAT3 and JAK3 levels, and decreasing various STAT3 downstream targets such as MCL-1, BCL-2, BCL-xL, and CYCLIN D3. Cucurbitacins' anti-cancer properties and underlying mechanisms in human stomach cancer remain unknown.

3.19 *Ephedra alata* Decne.

3.19.1 Background

The genus *Ephedra* is indigenous to the temperate and subtropical latitudes of Europe, Asia, and America, and grows especially in northern and western China, northern India, and Spain. In the United States, *Ephedra* plants grow along the Rocky Mountains.

Kingdom	**Plantae**
Division	Gnetophyta
Class	Magnoliopsida
Order	Cucurbitales
Family	Ephedraceae
Genus	*Ephedra*
Species	*alata*
Binomial name	*Ephedra alata* L. Decne.

3.19.2 Ethnomedical Considerations

E. alata is native to North Africa and Southwest Asia (Figure 3.19); it is distributed in Asia (Afghanistan, India, Punjab, Palestine, Kuwait, Qatar, Bahrain, Saudi Arabia, United Arab Emirates, Oman, and Yemen) and Africa. It is distributed Algeria, Egypt, Libyan, Morocco, Tunisia, Mauritania, Chad, Mali in Africa and Saudi Arabia, Iraq, Iran, Palestine, Lebanon, Jordan, and Syria in Asia.

3.19.3 Morphology and Pharmacognostic Characteristics

3.19.3.1 Morphology

E. alata is a short, evergreen shrub that can grow up to 2 ft (60–90 cm) in height. The stems are green, small, straight, or upright, with small ribs and grooves, about 1.5 mm wide, and usually end in a sharp spot. The nodes are 4–6 cm apart, and small triangular leaves emerge from the stem areas. The nodes are reddish brown. Graduates often form the basis. They bear little yellow flowers and fruits, and they give off a strong pine-like aroma. *Ephedra foliata* is a bad or consecutive plant. The branches are small.

3.19.3.2 Pharmacognostic Characteristics

A TS of the stem shows a single-layered epidermis composed of quadrangular cells with a thick-walled cuticle and papillae on the outer side. Vertical rows of sunken stomata are present along the sides of the furrows between the ridges. The cortex is composed of thin-walled, radially elongated, and loosely arranged chlorenchymatous

FIGURE 3.19 Photograph of *Ephedra alata*.

cells containing abundant chloroplasts (palisade cells) and intercellular spaces. Non-lignified fibres (hypodermal fibres) like a bunch of grapes occur below the ridges where no palisade cells are present. Lignified fibres (isolated or in groups) and micro prismatic crystals of calcium oxalate are also randomly found in the cortex. The innermost layer of the cortex is an ill-defined endodermis which surrounds a ring of collateral, conjoint, and open vascular bundles.

3.19.4 Phytochemistry

Preliminary phytochemical analysis of *E. alata* has shown the presence of cardiac glycosides, reducing sugars, flavonoids, phenolic compounds, and alkaloids. *Ephedra* varieties contain the alkaloids ephedrine, pseudoephedrine, norephedrine, norpseudoephedrine, methylephedrine, and methyl pseudoephedrine. Side E-type alkaloids, ephedroxane, and macrocyclic spermidines called ephedradine A–D are isolated from another Eurasian species *Ephedra*. The total number of alkaloids isolated from the air components of *E. alata* was 0.2– 0.22%. Flavonoids separated from *E. alata* included vicenin II, lucenin III, kaempferol 3-rhamnoside, quercetin 3-rhamnoside, herbacetin 7-glucoside, herbacetin 8-methyl ether 3-glucoside-7-*O*-rutinoside and herbacetin 7-*O*-(6″-quinylglucoside).

3.19.5 Pharmacological Effects

Ephedrine produced similar effects to epinephrine on smooth muscles. It blocked access to the abdominal tract and did not regenerate the splenic capsule and pilomotor muscles. It had the same actions on another myometrial and urinary tract as epinephrine. Ephedrine also produced hyperglycaemia and eosinopenia. Ephedrine has been implicated in muscle weakness of myasthenia gravis in an unknown way. Isolated skeletal muscle was detected by ephedrine; the observed effect was not dependent on ephedrine actions on the vascular system or CNS. Abortion is prevented by ephedrine. Ephedrine also reduces the release of pancreatic juice.

The antioxidant activity of *E. alata* was tested with 2,2-diphenyl-1-picryl-hydrazyl-hydrate assay. *E. alata* methanolic extract showed high antioxidant activity and the ability to release strong oxygen; without strong oxygen, the plant's IC_{50} was almost equal to the normal antioxidant Trolox.

3.19.6 Scientific Investigation for the Management of Cancer

The plant extract displayed a significant cytotoxic effect on MCF7 with IC_{50} and IC_{90} compared with doxorubicin as a positive control.

3.19.7 Reported Mechanism of Action as Anti-cancer Activity

The methanolic extract of *E. alata* showed a significant scavenging effect on the DPPH radical, compared to gallic acid. The results showed an honest correlation between antioxidant activity and flavonoid content. The plant extract displayed a significant cytotoxic effect on MCF7 with IC_{50} and IC_{90} compared with doxorubicin as a positive control. The extract of *E. alata* had significant antioxidant and cytotoxic effects in addition to their antimicrobial activity. The antioxidant effects are exerted by phenolic components, which act as free radical scavenger, oxygen radical absorbance, and metal ion chelators and interfere with the cytochrome P450 mixed-function oxidases.

3.19.8 Toxicological Aspects

According to a 2004 study by the Food and Drug Administration (FDA), dietary supplements containing the *Ephedra* alkaloid represented an unacceptable health risk. The FDA banned all ephedrine-containing drugs. Many cardiovascular events were associated with the use of dietary supplements containing the *Ephedra* alkaloids. Adverse events are reported in people who have taken dietary supplements containing ephedrine and related alkaloids (pseudoephedrine, norephedrine, and *N*-methyl ephedrine). The reports recorded adverse events ranging from seizures and headaches to death in eight ephedrine users and included reports of stroke, myocardial infarction, chest pain, fainting, insomnia, nausea and vomiting, fatigue, and dizziness.

3.19.9 Conclusion and Future Aspects

Additional studies must now be conducted on a wider panel of breast cancers of different stages and different grades in order to better understand the chemosensitization by *E. alata*, but also to determine if *E. alata* can act in synergy with other anti-cancer agents, by which it could enhance the activity by modulating one or more mechanisms of resistance.

3.20 *ARBUSTUS ANDRACHNE* L.

3.20.1 Background

Arbutus andrachne L. or "Greek Strawberry Tree," known in Arabic as Qyqab, is a member of the Ericaceae family. It is a small evergreen tree, stretching from the Eastern Mediterranean to the North Black Sea region (Figure 3.20). In Palestine, the plant grows on rocky hillsides, flowering from March to April, and the seeds ripen from September to October. The flowers are hermaphrodites and are pollinated by bees. The plant fertilizes itself. The red stems and evergreen leaves make the tree very attractive with a high value in the area. In addition, the fruit tastes good when ripe and can be eaten fresh, dried, or as jam.

Kingdom	**Plantae**
Division	Gnetophyta
Class	Magnoliopsida
Order	Ericales
Family	Ericaceae
Genus	*Arbutus*
Species	*andrachne*
Binomial name	*Arbutus andrachne* L.

FIGURE 3.20 Photograph of *Arbustus andrachne* L.

3.20.2 Ethnomedical Considerations

It is widely distributed throughout the Mediterranean region and Western Europe from north to west of France and Ireland. Because of its presence in south-western and north-western Ireland, it is known as the "Irish strawberry tree," or cain or cane apple (from the Irish name for the tree, caithne, or sometimes tree) of Killarney strawberries.

3.20.3 Morphology and Pharmacognostic Characteristics

3.20.3.1 Morphology

Arbutus andrachne is a remarkable tree with a peeling, red-brown bark and attractive white-cream spring flowers. The fruits are edible, red in colour, and give the common name of the tree as it resembles strawberries. *Arbutus* is native to the Mediterranean region.

3.20.4 Phytochemistry

The *andrachne* species was the highest among 51 other medicinal plants from Jordan with antioxidant content. According to the existing literature, this plant covers triterpenoids and steroids in the fruit. Triterpenoids, sterols, and lipids are separated from the bark, leaves, and fruits; arbutin, menotropins, and catechin are also separated from the bark and leaves. The current study focuses on catechin acquisition in different parts of *A. andrachne*.

Catechin is a basic unit of flavonoid and a small group of flavan-3-ols, representing the major c6-C3-C6 component of monomeric flavonoids C6-C3-C6, leading to a second plant use. Flavonoids are naturally occurring polyphenolic compounds that contain two benzene rings linked together by a pyrone ring or dihydropyrone ring flavonoids are usually glycosylated. Catechin contains two benzene rings called A and B rings and a dihydropyran heterocycle ring C with a hydroxyl group in carbon 3. There are two chiral sites in the carbon 2 and 3 molecules. Catechin consists of four diastereoisomers, two *trans*-isomers, a catechin activation, and two others in a secretion called *cis*-epicatechin. The most common catechin isomer is (+)-catechin and the common epicatechin is (−)-epicatechin; different epimers can be separated using chiral column chromatography. Catechin oxygenase, a key enzyme to the breakdown of catechin, is present in fungi and bacteria.

3.20.5 Pharmacological Properties

A. andrachne L. is a medicinal plant derived from the Jordan flora. It is of Mediterranean origin and is used to treat a wide range of medical conditions, including cancer, and have several reported therapeutic activities associated with anticancer activity. However, little is known about the antiproliferative effects of these plants on breast, colorectal, and skin cancers.

3.20.6 Scientific Investigation for the Management of Cancer

Strawberry tree (*Arbutus unedo* L.) honey (STH), also known as "bitter honey," is a typical product of the Mediterranean region. The main phenolic constituent of STH is homogentisic acid (2,5-dihydroxyphenylacetic acid [HGA]), which has remarkable antioxidant, antiradical, and protective properties against thermal cholesterol degradation, comparable to those of other, better-known antioxidants.

3.20.7 Reported Mechanism of Action as Anti-cancer Activity

Several studies have shown that STH has a very high content of phenolic compounds and strong antioxidant capacity. It also exerts antibacterial and anti-inflammatory activity with high anti-mutagenic and anti-proliferative properties important for tumour prevention and treatment.

3.20.8 Toxicological Aspects

It has been found that the plant exhibits low cytotoxicity. The plant is considered safe for its usage as an anti-cancer agent.

3.20.9 Conclusion and Future Aspects

The low cytotoxicity and high biocompatibility of strawberry tree honey, its extract, and homogentisic acid as its major component are a good starting point for further research into their biological effects on other models.

3.21 *EUPHORBIA HIEROSOLYMITANA* BOISS.

3.21.1 Background

Euphorbia is the largest and most diverse genus of flowering plant, commonly called spurge, in the family Euphorbiaceae. "Euphorbia" is sometimes used in general English to refer to all members of the Euphorbiaceae (with respect to genera), not just to members of the genus. This species has more than or about 2,000 members, making it one of the largest flowering species. It also has a wide range of chromosome calculations, as well as Rumex and Senecio. *Euphorbia antiquorum* is a species of the genus *Euphorbia*.

Kingdom	**Plantae**
Division	Tracheophyta
Class	Magnoliopsida
Order	Malpighiales
Family	Euphorbiaceae
Genus	*Euphorbia*
Species	*E. hierosolymitana*
Binomial name	*Euphorbia hierosolymitana* Boiss.

3.21.2 Ethnomedical Considerations

Euphorbia flowers are small, and a variety of pollinators, in different forms and colours, in the cyathium, involucre, cyathophyllum, or additional components such as glands are attached to these (Figure 3.21). The leafy spurges, *Euphorbia* subgenus *Esula*, have a very diverse habitat in the Mediterranean, considered to be one of 25 biodiversity hotspots. The eastern Mediterranean, in particular, is famous for its rich variety and is therefore considered a breeding ground.

FIGURE 3.21 Photograph of *Euphorbia hierosolymitana* Boiss.

3.21.3 Morphology and Pharmacognostic Characteristics

3.21.3.1 Morphology

Plants are seasonal trees, two-moving or long-lived, woody trees, or poisonous latex trees. The roots are right or thick and fleshy or tuberous. Many species are fragrant, thorny, or unarmed. The main trunk and hind limbs are thick and fleshy, 15–91 cm (6–36 in.) long. The leaves may twist, rotate, or whorl. In succulent species, the leaves are usually small and temporary. Stipules are very small and are partially converted to spines or glands, or non-existent.

3.21.3.2 Pharmacognostic Characteristics

E. hirta belongs to the plant family Euphorbiaceae and the genus *Euphorbia*. It is a slender-stemmed, annual hairy plant with many branches from the base to top, spreading up to 40 cm in height, reddish or purplish in colour. Leaves are opposite, elliptic-oblong to oblong-lanceolate, acute or subacute, dark green above, pale beneath, 1–2.5 cm long, blotched with purple in the middle, and toothed at the edge. The fruits are yellow, three-celled, hairy, keeled capsules, 1–2 mm in diameter, containing three brown, four-sided, angular, wrinkled seeds.

3.21.4 Phytochemistry

Euphorbia species contain methyl esters and derivatives as well as diterpene polyesters and other terpene compounds. The oil consists mainly of sesquiterpenes, and a small percentage of monoterpenes and aliphatic compounds. Eleinol (57.5%) was found to be the major constituent among the 24 compounds identified in *Euphorbia teheranica* Boiss. Other constituents found in *Euphorbia* species include ingenol 3-angelate, kaempferol 3-*O*-glucopyranoside, kaempferol, scopoletin, quercetin, vanillic acid, (*E*)-*p*-hydroxycinnamic acid, protocatechuic acid, 6,7-dihydroxy coumarin, beta-sitosterol, and daucosterol, jolkinolide B, 3,7,15-tri-*O*-acetyl-5-*O*-nicotinoyl-13,14-dihydroxymyrsinol, ellagic acid, 3,3′-di-*O*-methylellagic acid, 3,3′-di-*O*-methylellagic acid-4′-*O*-alpha-D-arabinfuranoside, 3,3′-di-*O*-methylellagic acid-4-*O*-beta-D-glucopyranoside, 3,3′, 4′-tri-*O*-methylellagic acid, 3-*O*-methylellagic acid-4′-*O*-beta-D-xylopyranoside, 3,3′,4-tri-*O*-methylellagic acid-4′-*O*-beta-D-glucopyranoside, brevifolin, 3,3′-di-*O*-methylellagic acid-4′-*O*-beta-D-xylopyranoside, and ethyl brevifolin carboxylate, piceatannol, and octacosyl *cis*-ferulate, octacosyl *trans*-ferulate, cholest-5-*en*-3-beta-ylhexadecanoate, chrysophanol, and octadecanoic acid.

3.21.5 Pharmacological Effects

Anti-cancer properties: In vitro studies indicate that *Euphorbia lagascae* seeds piceatannol and other methylated extracts have increased apoptosis in some cancer cell lines.

Antidiabetic Activities: In an *in vitro* study, *Euphorbia balsamifera* was a peroxisome proliferator-activated receptor.

Anti-eczema properties: Several studies indicate that *Euphorbia* has an irritating effect on the skin. In theory, this may be helpful in "burning" eczema and warts. Another clinical trial also reported that taking *Euphorbia acaulis* orally as a tablet may be helpful in treating eczema.

Anti-inflammatory properties: Ethanol extracts of the whole plant *Euphorbia prostrata* and its various components may show anti-inflammatory properties, as seen in carrageenan animal models. In addition, studies of acute inflammation of the components using histamine and bradykinin-induced pedal oedema have shown specific inhibition of histamine-induced oedema, which raises the initial phase of the acute inflammatory response.

Antioxidant properties: In animal studies, mice with endogenous polysaccharide treatment extracted from *Euphorbia kansui* had higher enzymatic activities of superoxide dismutase and glutathione peroxidase associated with a decrease in malondialdehyde dehydration after exercise.

Euphorbia is reported to have anti-plant properties. In 12 terpene fragments found in the roots of *Euphorbia kansui*, eight showed significant inhibition of cell proliferation at low concentrations in *Xenopus* embryo cells. Four terpenes blocked only cell proliferation at very high altitudes. It was also found that most of the diterpene computers blocking the proliferation of cell phones also inhibit the activity of topoisomerase II.

3.21.6 Scientific Investigation for the Management of Cancer

Euphorbia tirucalli L. significantly improved survival and at the same time reduced tumour growth in the peritoneal area in mice. It is thought that the modulatory effect of *E. tirucalli* L. on myelopoietic response and prostaglandin E levels may be related to its anti-tumour activity as a possible means of controlling granulocyte and macrophage production and expression of functional activity. Jolkinolide B, a component of *Euphorbia fischeriana*, is significantly reduced in the proliferation of three cancer cell lines, possibly by binding the cell cycle to the G1 stage and later forming apoptosis.

3.21.7 Reported Mechanism of Action as Anti-cancer Activity

The plant is able to arrest angiogenesis by increasing antioxidant enzymes and inhibiting reactive oxygen species production and oxidative stress. The sterols even block inflammatory cytokines and induced apoptosis.

3.21.8 Toxicological Aspects

It has been found that the plant exhibits low cytotoxicity. The plant is considered safe for its usage as an anti-cancer agent.

3.21.9 Conclusion and Future Aspects

E. cuneata as a natural pharmacological agent to replace chemical drugs could be a promising step. The use of *E. hierosolymitana* as a potential anti-cancer agent is therefore promising as the isolation of cytotoxic compounds from crude extracts and the use of such compounds can prevent or stall the progression of cancer.

3.22 *HYPERICUM PERFORATUM* L.

3.22.1 Background

The genus *Hypericum* is derived from the Greek words "hyper" (above) and "eikon" (pictured), referring to the traditional use of the plant to avoid evil by hanging plants on a religious icon in the house during St. Mary's Day. *H. perforatum* originates in the temperate regions of Europe and Asia, but it has spread to temperate regions around the world like a weed invading the entire globe. It was introduced to North America from Europe. The flower grows from the plains, the pastures, and the disturbed fields. It prefers sandy soils.

Kingdom	**Plantae**
Division	Tracheophyta
Class	Magnoliopsida
Order	Malpighiales
Family	Hypericaceae
Genus	*Hypericum*
Species	*perforatum*
Binomial name	*Hypericum perforatum* L.

3.22.2 Ethnomedical Considerations

St. John's wort (SJW), botanically known as *H. perforatum*, is a sprawling leafy herb that grows in open, disturbed areas across much of the world's temperate regions (Figure 3.22). It produces dozens of biologically active substances, although two hypericin and hyperforin (lipophilic phloroglucinol) have the greatest medical activity. Other compounds, including the flavonoids rutin, quercetin, and kaempferol, also appear to have medical activity. *H. perforatum* has been intensively studied on isolated tissue specimens, using animal models and in human clinical trials.

FIGURE 3.22 Flower of St. John's wort.

3.22.3 Pharmacognostic Characteristics

3.22.3.1 Morphology

H. perforatum consists of freely branching shrubby herbs that typically are 40–80 cm in height. The stems and branches are densely covered by oblong, smooth-margined leaves that are 1–3 cm long and 0.3–1.0 cm wide. The leaves are interrupted by minute translucent spots that are evident when held up to the light. The upper portions of mature plants can produce several dozen five-petalled yellow flowers that are typically 1–2 cm wide.

3.22.4 Phytochemistry

Chemical research has found seven groups of compounds that work with the drug in *H. perforatum*. They include naphthodianthrones, phloroglucinols, flavonoids (such as phenylpropanoids, flavonol glycosides, and biflavones), and essential oils. There are two major active components identified: hypericin (naphtho dianthrone) and hyperforin (phloroglucinol). Flavonoids in SJW range from 7% on stems to 12% on flowers and leaves. Flavonoids include flavonols (kaempferol, quercetin), flavones (luteolin), glycosides (hyperside, isoquercitrin, and rutin), biflavones (biapigenin), amentoflavone, myricetin, hyperin, oligomeric proanthocyanidins, and miquelianin, all biogenetically related. SJW quotes contain several classes of lipophilic compounds with a prescribed therapeutic value, including the extracts of phloroglucinol and oils. These include tannins (3–16%), phenolic compounds (caffeic acid, chlorogenic acid, and coumaric acid), xanthones, and hyperfolin. Additional compounds include acids (nicotinic, palmitic, and stearic) pectin, hydrocarbons, carotenoids, choline, and alcohol chain chains.

3.22.5 Pharmacological Effects

Hyperforin and *hypericin* have also been tested for their anti-cancer properties. Hyperforin inhibits tumour cell growth *in vitro*. Hypericin being a photodynamic compound photoactivated by white light or ultraviolet light or both can cause almost complete apoptosis (94%) in malignant T cells of the skin and lymphoma T cells. Laser irradiation has led to toxic effects on human prostatic cancer cells, bladder carcinoma cells, and lines of pancreatic cancer cells in *in vitro* systems.

As a matter of concern, hypericin alone has a weak effect on inhibiting cancer cell growth, while the methanolic release of SJW and hypericin leads to long-term inhibition of cell growth, induces apoptosis, and reduces phototoxicity toxicity. Considering the stimulant effects of hyperforin and hypericin as anti-cancer agents, further research is needed to evaluate their efficacy, mechanism of action, and adverse reactions.

3.22.6 Scientific Investigation for the Management of Cancer

H. perforatum L. also known as St. John's wort is known to have many beneficial properties for the organism, including its antioxidant and anti-cancer activities. It is also known to have shown anti-proliferative and cytotoxic effects against various cancer cell lines.

3.22.7 Reported Mechanism of Action as Anti-cancer Activity

Due to high lipophilicity and mild acidity, *H. perforatum* L. accumulates in membranes and acts as a protonophore that hinders inner mitochondrial membrane hyperpolarization, inhibiting mitochondrial ROS generation and consequently tumour cell proliferation. At the plasma membrane level, HPF prevents cytosol alkalization and extracellular acidification by allowing protons to re-enter the cells. These effects can revert or at least attenuate cancer cell phenotype, contributing to hamper proliferation, neo-angiogenesis, and metastatic dissemination.

Hyperforin and hypericin have also been examined for their anti-cancer properties. According to Schempp et al. (2002), hyperforin inhibits tumour cell growth *in vitro*. The mechanism involves the induction of apoptosis (programmed cell death) through the activation of caspases, which are cysteine proteases that trigger a cascade of proteolytic cleavage occurrences in mammalian cells. Hyperforin also causes the release of cytochrome c from isolated mitochondria. A mitochondrial activation is an early event in hyperforin-mediated apoptosis, and hyperforin inhibits tumour growth *in vivo*.

3.22.8 Toxicological Aspects

The *H. perforatum* extract used as a standardized and titrated extract in flavonoids, hyperforin, and hypericin is a safe and efficacious therapy for the treatment of cancer.

3.22.9 Conclusion and Future Aspects

H. perforatum L. is a promising chemopreventive agent and further studies are needed in order to evaluate the full potential of this plant.

3.23 *CROCUS SATIVUS* L.

3.23.1 Background

C. sativus L. is a small perennial plant, considered the king of the spice world. It belongs to the Iridaceae family. *Crocus* genus contains about 90 species and some are cultivated for flowering. *C. sativus* contains three major components: crocin, picrocrocin, and safranal. These three components are reportedly responsible for the colour, flavour, and aroma of saffron.

The saffron plant belongs to the Iridaceae family. This herbaceous perennial plant reaches 10–to 25 cm in height from its bulbs. The bulb, of sub-ovoid shape, has a dynamic size and form. It has a large structure and is covered with many spates. Each mother bulb produces one to three adult apical buds and a few small lanterns from lateral buds. Saffron has two types of roots: fibrous and thin roots at the base of the mother globe, and symmetrical roots formed at the base of the stems. The leaves vary from 5 to 11 shoots each. They are very small and measure 1.5–2.5 mm in size. They measure 20–60 cm long with a white band on the inside and a rib on the outside.

Kingdom	**Plantae**
Division	Tracheophyta
Class	Magnoliopsida
Order	Asparagales
Family	Iridaceae
Genus	*Crocus*
Species	*sativus*
Binomial name	*Crocus sativus* L.

3.23.2 Ethnomedical Considerations

Saffron is one of the most expensive cash crops among medicinal plants in the world and thus it has been called "the red gold." It has been known for more than 4,000 years and was used mostly in traditional medicine as a tonic agent and antidepressant drug. Saffron is obtained from the dried red stigmas of *C. sativus* L., an autumnal herbaceous flowering plant belonging to the Iridaceae family (Figure 3.23). It is largely cultivated in Iran, India, Afghanistan, Greece, Morocco, Spain, and Italy.

FIGURE 3.23 Flower of *Crocus sativus* L.

3.23.3 Pharmacognostic Characteristics

3.23.3.1 Morphology

Colour—reddish-black to reddish-brown. The style changes from yellow to orange-yellow. Smell—strong, characteristic, and delicious. Taste—feature, and bitterness. Size—stigmas 25 mm long, and styles are approximately 10 mm long. Shape—stigma trifid and cylinder styles.

3.23.3.2 Microscopic Characteristics

If the immersed tree is inspected under a lens or a microscope, the spots will be found to be different or combined three or more times in the yellow styles. Each stem is about 25 mm long and has the shape of a thin funnel, with rough edges or fimbricate.

3.23.4 Phytochemistry

The ingredients of saffron are crocin (colour-coded), picrocrocin (anti-flavour), and safranal (anti-scent). Sapphire contains more than 150 flexible ingredients and fragrances. It also contains many active components, many of which are carotenoids, including zeaxanthin, lycopene, and α- and β-carotene. Fragrant computers are associated with more than 34 substances, namely terpenes, terpene alcohols, and their esters. Non-volatiles include 14 crocins that are responsible for the red colour of stigmas as well as carotenes, crocetin, picrocrocin, a bitter substance and the safranal main regimen.

3.23.5 Pharmacological Effects

According to reports, from earlier times, the healing power of *C. sativus* is well documented. In this case, saffron water extract (carotenoid) has been shown to be effective in treating cancer and cerebrovascular and heart problems. It is also reported that there are other activities in different parts of the world. In the Middle East, reportedly used as an antispasmodic, aphrodisiac, carminative, cognition enhancer, emmenagogue and thymoleptic in traditional Chinese medicine, saffron was used to treat amenorrhea, severe delivery, menorrhagia, and postpartum lochiostasis. In the Indian medicine system, saffron was used to treat bronchitis, fever, headaches, sore throat, and vomiting. Various medical activities are also reported on saffron, i.e. antihypertensive activity, anticonvulsant activity, antitussive activity, and the anti-inflammatory action of saffron antioxidant activity. In addition to medicinal value, saffron is used as a spice (regarded as the permanent king of the spice world), as a dye, and a perfume in the food industry.

3.23.6 Scientific Investigation for the Management of Cancer

Chemoprevention using readily available natural substances from vegetables, fruits, herbs, and spices is one of the significantly important approaches for cancer prevention in the present era. Among the spices, *C. sativus* L. has generated interest because pharmacological experiments have established numerous beneficial properties, including radical scavenging, anti-mutagenic and immuno-modulating effects. The more powerful components of saffron are crocin, crocetin, and safranal. Studies in animal models and with cultured human malignant cell lines have demonstrated the anti-tumour and cancer-preventive activities of saffron and its main ingredients.

3.23.7 Reported Mechanism of Action as Anti-cancer Activity

The mechanisms underlying cancer chemopreventive activities of carotenoids include modulation of carcinogen metabolism, regulation of cell growth and cell cycle progression, inhibition of cell proliferation, antioxidant activity, immune modulation, enhancement of cell differentiation, and stimulation of cell-to-cell gap junction communication, apoptosis, and retinoid-dependent signalling. Various hypotheses for the anti-tumour actions of saffron and its components have been proposed: (1) the inhibitory effect on cellular DNA and RNA synthesis, but not on protein synthesis; (2) the inhibitory effect on free radical chain reactions; (3) the metabolic conversion of naturally occurring carotenoids to retinoids; and (4) the interaction of carotenoids with topoisomerase II, an enzyme involved in cellular DNA–protein interaction. Additionally, the immunomodulatory activity on driving towards T helper cell type 1 (Th1) and Th2 limbs of the immune system of saffron was also demonstrated.

3.23.8 Toxicological Aspects

Saffron, a spice obtained from the flower of *C. sativus*, is rich in carotenoids. Two major natural carotenoids of saffron, crocin, and crocetin, are responsible for their colour. Preclinical evidence has demonstrated that dietary intake of some carotenoids has potent anti-tumour effects both *in vitro* and *in vivo*, signifying their potential preventive and/or therapeutic roles in several tissues.

3.23.9 Conclusion and Future Aspects

Given its therapeutic and economic importance, its natural abundance, in addition to its common usage in ethnic medicine, saffron provides a varied and accessible platform for phytochemical-based drug discovery. A consolidation of its traditional usage as well as its chemical and pharmacological profiles will thus guide efforts aimed at maximizing this potential. A stronger focus on clinical studies and phytochemical definition will be essential for future research efforts. Hence, it is suggested that future research be warranted that will define the possible use of saffron as an effective anti-cancer and chemopreventive agent in clinical trials. Although data from *in vitro* studies look convincing, well-designed clinical trials in humans are needed to ascertain whether saffron can become part of our armamentarium against cancer.

3.24 *MELISSA OFFICINALIS* L.

3.24.1 Background

M. officinalis is native to New Zealand and recognized as a bush balm. Lemon balm is a strong perennial herb with a variety of dental leaves, square ovate, and branch stems having a tree-like appearance, its height can range from less than 8 in. to about 5 ft, and is 12–24 in. wide. The leaves may be smooth and the fruit is small nuts. The small flowers of lemon balm have two lips, grow into dense clusters, and can be pale yellow, white, red, and rarely purple or blue, and have non-glandular hair.

The plant predominates in the eastern Mediterranean province and West Asia. Backyard farming of lemon balm is well-liked in Europe. The plant is found growing wild in sunny places on elevations as much as about 1,000 m.

Kingdom	**Plantae**
Division	Tracheophyta
Class	Magnoliopsida
Order	Lamiales
Family	Lamiacaeae
Genus	*Melissa*
Species	*officinalis*
Binomial name	*Melissa officinalis* L.

3.24.2 Ethnomedical Considerations

M. officinalis is widely cultivated in some parts of Iran. The leaves of *M. officinalis* L. (Lamiaceae) are used in Iranian folk medicine for their sedative, tonic, carminative, digestive, antispasmodic, analgesic, and diuretic properties, as well as for functional gastrointestinal disorders. The perennial herb *M. officinalis* is also called a lemon balm, bee balm, or honey balm. This plant is an associate of the Lamiaceae (mint) family and belongs to a genus that includes five species of recurrent herbs resident in Europe, Central Asia, and Iran. Even though *M. officinalis* was initiated primarily in Southern Europe, it is now naturalized around the world, from North America to New Zealand (Figure 3.24).

3.24.3 Morphology and Pharmacognostic Characteristics

3.24.3.1 Morphology

M. officinalis (lemon balm) is an enduring herbaceous plant in the mint family Lamiaceae. Lemon grass is a tall plant having enormous striped leaves with an uneven edge. The plants live for a long time; the yield plant is supplanted following five years to permit the ground to restore. Lemon salve seeds require light and a base temperature of 20°C (68°F) to grow. The stems stand outwards. The leaves are straightforward (i.e., lobed or unlobed yet not isolated into leaflets). The heart-moulded leaves are 2–8 cm (0.79–3.15 in.) long and have a harsh, veined surface. They are delicate and furry with scalloped edges and have a gentle lemon fragrance.

FIGURE 3.24 Photograph of *Melissa officinalis* L.

During summer, little white or pale pink blossoms show up. Bloom length fluctuates between 8 mm and 15 mm and it has a prevalent ovary and comes up short on hypanthium. The seeds are tiny, around 1–1.5 mm long, with a dull brown or dark tone. The dry natural product parts into areas, each holding at least one seed yet doesn't part open when ready. The plant has a charming smell, for instance, anise, natural product, mint, or resin.

3.24.3.2 Pharmacognostic Characteristics

The microscopic study of TS of stem and root of *M. officinalis* shows that the upper part of the stem consists of an outer layer of skin, followed by a narrow cortex of subepidermal cells and parenchymatous cells. The phloem includes a little ring of slim-walled cellulosic sifter cylinders and friend cells related to phloem parenchyma. No lignified components are available. The xylem comprises lignified xylem vessels with winding thickening, wood strands, and wood parenchyma. The lower portion of the stem contains an external epidermis conveying non-glandular trichomes, notwithstanding a couple of quantities of glandular hairs, trailed by a limited cortex, which is separated into collenchyma and parenchyma. The root is almost roundabout in the frame, showing an outside layer of earthy-coloured stopper cells encompassing a tight cortex. The vascular tissue comprises a phloem framed of delicate cellulosic components enveloping strainer cylinders and adjoining cells, joined by phloem parenchyma and phloem fibre alongside a wide xylary locale stretching out to the middle.

3.24.4 Phytochemistry

Lemon balm includes eugenol, tannins, and terpenes. It also contains (+)-citronellal, 1-octen-3-ol, 10-α-cadinol, 3-octanol, 3-octanone, α-cubebene, α-humulene, caffeic acid, caryophyllene, caryophyllene oxide, catechin, chlorogenic acid, *cis*-3-hexenol, *cis*-ocimene, citral A, citral B, copaene, δ-cadinene, eugenyl acetate, γ-cadinene, geranial, geraniol, geranyl acetate, germacrene D, isogeranial, linalool, luteolin-7-glucoside, methyl heptenone, neral, nerol, octyl benzoate, oleanolic acid, pomolic acid ((1*R*)-hydroxyursolic acid), protocatechuic acid, rhamnazin, rosmarinic acid, stachyose, succinic acid, thymol, *trans*-ocimene, and ursolic acid.

3.24.5 Pharmacological Properties

M. officinalis L. is a notable restorative plant. The plant has been utilized for the therapy of mental and CNS infections, cardiovascular and respiratory issues, and different tumours, and as a memory enhancer, heart tonic, antidote, tranquilizer, and wound sanitizer. Restorative exercises of *M. officinalis* are viewed as credited primarily to its medicinal balm (EO) and phenolic compounds. *M. officinalis* has been broadly utilized for the therapy of a few kinds of diseases. The significant part *of M. officinalis*, citral, has been displayed to initiate apoptosis of GBM cell lines that communicated dynamic MRP1. Citral has likewise been compelling in lessening the reasonability of a few human cancer cell lines and a mouse melanoma cell line. The apoptosis-prompting exercises of the plant are potentially interceded by EO parts too as lipophilic constituents which can interface with the cell layer and pass through it. Certain phenolic acids, which are generous in polar concentrates, represent the antimutagenic and antigenotoxic impacts of *M. officinalis*.

3.24.6 Scientific Investigation for the Management of Cancer

In 2015, Javadi et al. remarked that lemon balm has shown promising outcomes in a few kinds of carcinoma. Weidner et al. have demonstrated that the hydroalcoholic extracts applied anti-proliferative consequences for colon carcinoma cells and actuated apoptosis by means of enlistment of intracellular ROS age. A 50% ethanol concentrate of *M. officinalis* has been accounted for cytotoxicity that affect human colon disease cell lines as indicated by MTT and NR tests that are cell practicality measures frequently used to decide cytotoxicity following openness to harmful substances. Carocho et al., in 2015, dealt with the decoctions of the plant containing RA and lithospermic acid A and uncovered development hindrance action against various human growth cell lines.

3.24.7 Reported Mechanism of Action as Anti-cancer Activity

The dichloromethane part of *M. officinalis* fundamentally prompted apoptosis in leukaemia cell lines by means of up-guideline of Fas and Bax mRNA articulation and expanding the Bax/Bcl-2 proportion, demonstrating its ability in enacting both extraneous and natural pathways of apoptosis. The apoptosis-inciting activities of the plant are perhaps intervened by EO parts too as lipophilic constituents which can connect with the cell layer and pass through it.

3.24.8 Toxicological Aspects

M. officinalis is a strong anti-carcinoma agent that can be used to take care of cancer cells. *M. officinalis* extract is a safe herb, which can be used as a tea. However, further *in vivo* and clinical studies are needed to assess the practical health effect of *Melissa* extracts in cancer treatment.

3.24.9 Conclusion and Future Aspects

Further *in vivo* and experimental studies are needed to assess the practical health effect of *Melissa* extracts in cancer treatment. Besides, mechanisms of actions and pharmacokinetics of various active compounds of *M. officinalis* extract should be elucidated to reach a realistic dosage in cancer treatment.

3.25 *Origanum jordanicum*

3.25.1 Background

O. jordanicum L. belongs to the Lamiaceae family. It is a perennial frostbite, native to both Cyprus the Mediterranean Basin, especially in the cooler Himalayas. Most of these species are fragrant and grow wild in the Mediterranean region. This genus contains more than 44 species, 6 subspecies, 3 botanical varieties, and 18 natural occurring hybrids. It also includes a few species of oregano and sweet marjoram.

It is commonly known as the sweet marjoram and originates in Cyprus, Anatolia (Turkey), and is also native to parts of the Mediterranean region and especially to Egypt. It is grown all over the world: in different parts of India, France, Hungary, and the United States because of its delicious taste and aroma. Marjoram was originally used by Hippocrates as an antiseptic agent.

Kingdom	**Plantae**
Division	Tracheophyta
Class	Magnoliopsida
Order	Lamiales
Family	Lamiaceae
Genus	*Origanum*
Species	*jordanicum*
Binomial name	*Origanum jordanicum*

3.25.2 Ethnomedical Considerations

Origanum is one of 200 genera of the Lamiaceae family, containing 3,500 species worldwide (Figure 3.25). The vast majorities of species is fragrant and grow naturally in the Mediterranean region. The genus is characterized by large morphological and chemical diversity. *Origanum* is a monospecific segment broadly disseminated in North Africa and in temperate and dried zones of Eurasia.

FIGURE 3.25 Photograph of *Origanum jordanicum*.

3.25.3 Pharmacognostic Characteristics

3.25.3.1 Morphology

Origanum majorana L. is a dense hardy perennial tree. It has a square stem with many red branches spilling to form a mound. The straight stems become weak, hairy, round, and green with red dots. The leaves are smooth, light, petiolate, and ovate to oval-green, grey-green in colour arranged opposite each other in a square stem. The texture is extremely smooth due to the presence of many hairs. They are 0.5–1.5 cm long and 0.2–0.8 cm wide, with obtuse apex, total border, a parallel but narrower base, and reticulate venation 16. Marjoram has a small, double-lipped, tube, white or soft pink blossoms, and green bracts that flourish in clusters from mid- to late summer (June to September). They are <0.3 cm long and are arranged with 1.3 cm long burr-shaped heads. The flowers are hermaphrodites naturally. The seeds are mint, oval, black, and brown in colour and ripen from August to September. It has cylindrical tap roots, shrivelled with opposite cracks: 0.2–0.6 mm wide. The outside of the root is dark brown while the light brown inside has several long roots and root scars. Fractures are long, unusual, and have fragrant and bitter fragments.

3.25.3.2 Microscopic Characteristics

Diacytic type stomata are present in the leaves, which are evenly distributed by the presence of veins, vein islets, and upper extremity veins. Polygonal cells are present in the upper epidermis while most cover the trichomes on the outside. The cover trichomes are multicellular, uniseriate, sharp, and have small walls.

The leaf exhibits a cuticularized epidermis that comprises layers of coarse collenchyma followed by vascular bundles, while mesophyll showed only palisade cells and sponge parenchyma. Collenchyma tissues consist of thick circular parenchyma cells and 16 xylem fibres. The stem of the remedy is circular in the opposite part consisting of a thick cuticle. The epidermis is fabricated from a lone layer of rectangular cells and five to six layers of adjacent polygonal parenchyma cells that form the cortex. Phloem and phloem parenchyma fibres are clearly separated.

Two cell-thick medullary beams alongside xylem vessels, xylem parenchyma, and noticeable parenchymatous substance are in participation in the middle. The root segment is round in frame consisting of two to three layers of rectangular plug cells with six to seven layers of personally packed parenchyma shaping the cortex. The xylem comprises xylem vessels and xylem parenchyma. The medullary ray is composed of two cell-thick rectangular cells. The phloem is outside the xylem, with no pith.

3.25.4 Phytochemistry

Sweet marjoram is characterized by a strong, spicy, and pleasant scent. Investigation of spice announced the presence of particularly unstable oil as significant constituents, inferable from its fragrant nature. Different phytochemical tests found the presence of flavonoids, terpenoids, and tannins in ethanol extricate, though saponins and sugars were available in stem and root water extract separately. Both the extracts have not shown the presence of glycosides, proteins, and alkaloids. Essential oil of the plant contains terpinen-4-ol (31.15%), *cis*-sabinene hydrate (15.76%), *p*-cymene (6.83%), sabinene (6.91%), *trans*-sabinene hydrate (3.86%), and α-terpineol (3.71%) as the core components. The majorly famous constituents of *O. majorana* are carvacrol (65%) and thymol (4%).

3.25.5 Pharmacological Effects

3.25.5.1 Antioxidant Activity

Ethanol extracted from marjoram leaves has shown antioxidant and free radical scavenging activity using colorimetric tests showing marked inhibitory effect on 1,1-diphenyl-2-picryl hydrazyl (DPPH) scavenging access. Ethanolic extracts of stems and roots have shown *in vitro* antioxidant activity, respectively using the spectrophotometric method by DPPH, H_2O_2 free radical scavenging, metal chelating, and ferric-reducing power assay. Both the extracts showed strong antioxidant activity overall. The IC_{50} values were comparable with ascorbic acid and the reducing ability of root ethanol extract was established to be high compared to stem ethanol extract.

3.25.6 Scientific Investigation for the Management of Cancer

Phytochemical analysis conducted by several scientists has shown that marjoram is rich in bioactive compounds with anti-cancer activity. It has been found to reduce the viability of various cancer cells, including colon, liver, and breast cancer cells. Luteolin triggers both intrinsic and extrinsic apoptotic pathways in various human cancer cells. Another important component of the herb, β-caryophyllene, emerges as a natural compound with anti-cancer potential. It has been reported to have anti-cancer potential.

3.25.7 Reported Mechanism of Action as Anti-cancer Activity

In 2005, Vagi reported that one of the most well-known benefits of marjoram essential oil due to phenolic compounds such as carnosic acid and ursolic acid is its antioxidant properties. It helps eliminate free radicals and repair damage caused by cellular metabolism and oxidative processes. It is reported on luteolin as an important phytochemical with anti-cancer potential.

3.25.8 Toxicological Aspects

Regarding safety, marjoram has no adverse effects on humans or animals and can be considered relatively safe and reported to be safe.

3.25.9 Conclusion and Future Aspects

O. majorana has been reported to have anti-invasive and anti-metastatic effects against highly proliferative and invasive human cancer cell lines. *O. majorana* has emerged as a promising chemopreventive and therapeutic candidate to inhibit breast cancer growth and metastasis by modulating the expression and activity of multiple target molecules. Due to today's growing interest in combination therapy using multiple anti-cancer agents affecting multiple targets/pathways, *O. majorana* has made further progress to identify new compounds in breast and colon cancers.

3.26 *ROSMARINUS OFFICINALIS* L.

3.26.1 Background

R. officinalis is better known as rosemary. Rosemary is indigenous to the Mediterranean province, from where it was introduced into all continents. It is sporadically cultivated as a decorative plant in the hilly regions of Java. In the Philippines, it is at present developed on a miniature range for the fresh-herb market. Rosemary is commonly propagated by cuttings, division, or by air layering. Seeds are sometimes used, but they are produced only under very favourable growing conditions and often only 10–20% of the seeds germinate. Transplanting to the field is done at a spacing of 45 cm between plants in rows 1.2 m apart. It is also common to produce rosemary in containers in greenhouses.

Kingdom	**Plantae**
Division	Tracheophyta
Class	Magnoliopsida
Order	Lamiales
Family	Lamiaceae
Genus	*Rosmarinus*
Species	*officinalis*
Binomial name	*Rosmarinus officinalis* L.

3.26.2 Ethnomedical Considerations

R. officinalis L. (rosemary) is a medicinal plant native to the Mediterranean region and cultivated around the world. It is constituted by bioactive molecules, the phytocompounds, responsible for implementing several pharmacological activities, such as anti-inflammatory, antioxidant, antimicrobial, anti-proliferative, anti-tumour, and protective, inhibitory, and attenuating activities. However, it could be found all over the world. It is a perennial and aromatic plant, shrub-shaped with branches full of leaves, having a height of up to 2 m, and green leaves that exude a characteristic fragrance. *R. officinalis* might be utilized as a flavour in cooking, as a characteristic additive in the food business, and as a decorative and restorative plant (Figure 3.26).

3.26.3 Pharmacognostic Characteristics

3.26.3.1 Morphology

R. officinalis has inverse leaves and "squarish" stems—however, not as unmistakably so as those of the genus *Salvia*. Its leaves are evergreen, direct, and have smooth edges that by and large will more often than not twist somewhat under the leaf's

FIGURE 3.26 Flower of *Rosmarinus officinalis*.

sharp edge. In view of the closeness in appearance to different evergreens, the leaves are frequently alluded to as "needles." The flowers are essentially bilabiate, with a much bigger lower lip. The lower lip is frequently set apart with particular spots and lines and makes an extraordinary landing cushion for honey bees—the bloom's pollinators. Two stamens curve up and under the upper lip (two conjoined petals) and a while later outward from the bloom with the pistil likewise situated and between the stamens. The other two of the sprout's five petals stick out and down from the sides of the blossom. The blossoms will generally frame in bunches at the leaf axils. The colour of the flower for the most part is a shade of delicate to medium blue or lavender blue; however, there are a few assortments with white or pinkish bloom shades. In development propensity, the plant can either be prostrate and rambling or standing and shrubby or some in the middle between. By and large, prostate assortments will quite often sprout significantly more than standing assortments. Seeds are little nutlets and each bloom can deliver up to four seeds.

3.26.3.2 Microscopic Characteristics

The microscopic study of TS of leaves of *R. officinalis* shows the following features. A slim waxy cuticle is available on the upper and lower surface of the leaf. The leaves bear non-glandular and glandular trichomes on the two sides of the lamina. Non-glandular trichomes are available on the veins and leaf edges. The leaf inside contains mesophyll parenchyma cells of two kinds: (1) palisade mesophyll, oval cells close to the upper epidermis organized opposite to the epidermal cells, and (2) light mesophyll. The pressed palisade mesophyll cells contain chloroplasts that are significant leaf sites for photosynthesis.

3.26.4 Phytochemistry

By varying the concentration of these molecules in each specimen of the plant, several plant compounds with pharmacological activity can be isolated from the essential oils and extracts of rosemary. The most commonly reported plant compounds are caffeic acid, carnosic acid, chlorogenic acid, monomeric acid, oleanolic acid, rosmarinic acid, ursolic acid, alpha pinene, camphor, carnosol, eucalyptus, rosmagial, rosmanol, rosmaquinone A and B, and decohinoki for decohinoki eugenol and luteolin.

3.26.5 Pharmacological Effects

3.26.5.1 Antioxidant Activity

The antioxidative action of rosemary extracts has been evaluated using different solvents. In this view, Inatani et al. reported that rosmanol showed four times the antioxidant capacity of BRT and BRA (synthetic antioxidants) in both linoleic acid and dosage. In addition, this research accounted for the antioxidant action of carnosol and rosmanol by TBA and ferric thiocyanate. They accounted for a relationship between activity and chemical structure as an antioxidant. Aruoma et al. studied the antioxidant and pro-oxidant potentials of rosemary.

3.26.5.2 Anti-proliferation Work

Raw ethanolic rosemary extracts have different effects on the increase in human leukaemia and breast cancer cells.

3.26.5.3 Cancer-fighting Activity

Numerous studies have reported on the anti-cancer processes of *R. officinalis*. Rosemary has shown significant anti-growth functions against multiple human cancer cell lines. Important compounds in plant products such as carnosic acid, carnosol, and rosmarinic acid have been shown to induce apoptosis in these cancer cells, probably through the production of nitric oxide. Carnosic acid emerges to be a powerful promoter of apoptosis. Rosemary extract too has an interesting anti-tumourigenic action. These anti-inflammatory and anti-tumourigenic activities of *R. officinalis* can be used to treat cancer in the future.

3.26.6 Scientific Investigation for the Management of Cancer

Elective medicines for neoplastic infections with new medications are important in light of the fact that the clinical adequacy of chemotherapy is many times diminished by security impacts. A few normal substances of plant beginning have been exhibited to find lasting success in the counteraction and treatment of various growths. *R. officinalis* L. is a herb that is developed in different regions of the world, expanding consideration is being coordinated towards the drug limits of rosemary, used for its calming, against the infective or anti-cancer activity.

3.26.7 Reported Mechanism of Action as Anti-cancer Activity

The anti-tumour impact of rosemary has been connected with different systems, for example, the cell reinforcement impact, antiangiogenic properties, epigenetic activities, guideline of the resistant reaction and mitigating reaction, adjustment of explicit metabolic pathways, and expanded articulation of onco-suppressor qualities.

3.26.8 Toxicological Aspects

In fact, the effects of rosemary have been studied in many cancer cell lines, but the concentrations used in *in vitro* experimentations are widely variable. This high variability leads to the need for a more systematic analysis to detect efficient RE doses *in vivo*.

3.26.9 Conclusion and Future Aspects

All in all, even though the utilization of rosemary and its subordinates in the treatment of neoplasms is a captivating field of study, enormous and controlled examinations should be led to conclusively explain the genuine effect of this substance in clinical practice.

3.27 *PLANTAGO LANCEOLATA* L.

3.27.1 Background

P. lanceolata is a variety of blossoming plants in the plantain. The perennial plant grows in dry meadows, fields, pastures, roadsides, banks, and waste places, preferring dry sandy soil. The plants are cut all through the budding season and used fresh, as juice, or dried for decoctions. The leaves are cut before flowering and dried.

Kingdom	**Plantae**
Division	Tracheophyta
Class	Magnoliopsida
Order	Lamiales
Family	Plantaginaceae
Genus	*Plantago*
Species	*lanceolata*
Binomial name	*Plantago lanceolata* L.

3.27.2 Ethnomedical Considerations

P. lanceolata L. is a well-known species of the genus *Plantago*. It is broadly cultivated in fields, side-of-the-road strips, fields, and green regions in the calm world 800 m above the ocean level. It has been used for therapeutic purposes to treat diseases such as wound-healing, inflammation, cancer, respiratory system disorder, blood circulation, reproductive system, and digestive organs. It has various applications as cosmetics, as metal removal from polluted areas, as an additive in foods, and as an insecticide. The extracts of the plant also showed different properties such as antioxidant, antibacterial, anti-inflammatory, rheological, and viscoelastic. Phytochemicals in the root, leaf, and seed of *P. lanceolata* L. incorporate iridoid glycosides, polyphenols, polysaccharides, and flavonoids, which have restorative potential (Figure 3.27).

3.27.3 Pharmacognostic Characteristics

This plant is a perennial plant that forms a rosette, with flowering stems that are leafless, silky, and hairy (10–40 cm or 3.9–15.7 in.). The lanceolate basal leaves are broad or straight, with no teeth but with three to five hard veins that line into a short petiole. The flower stem has deep grooves, ending with an ovoid inflorescence of a lot of tiny flowers each with a sharp bract. Each flower can produce up to 200 seeds. The flowers are 4 mm (0.16 in.) in size (green calyx, corolla brownish) and has four curved posterior lobes with chocolate-coloured midribs and extended white stamen. It is found in cool Eurasia, widespread throughout the British Isles, but is not found in highly acidic soils (pH <4.5). It is also widespread in the United States and Australia as a genre introduced.

The seeds are oblong on the side and are dark brown. The plant grows along the way. Its invisible flower head, from which emerges small flowers with soft stamens, looks like a spikelet at the last part of a long grass stalk and almost disappears among flowers of various colours. At the bottom, however, small lance-shaped leaves

FIGURE 3.27 Photograph of *plantago lanceolata* L.

give the plant a certain epithet lanceolate, forming a large rosette. The veins in the 20–40 cm long leaves have no branches like other plants but go with the length of the leaves. It blooms from May to September.

3.27.4 Phytochemistry

The phytochemical screening on dried leaves of *P. lanceolata* has affirmed the presence of saponins, tannins, alkaloids, steroids, terpenoids, flavonoids, and phenolic compounds. Various convergences of bioactive mixtures like flavonoids, coumarins lipids, and cinnamic acids and tannins are tracked down in the entire or isolated pieces of *P. lanceolata* L. like flowers, leaves, and roots. Some significant phenolic intensities like 3,4-dihydroxyphenylacetic corrosive, (+)-catechin, kaempferol, and bioactive like gallic acid have been accounted for in the plant. A couple of flavonoid bioactive substances, including luteolin-7-*O*-glucuronide, luteolin, apigenin, luteolin-7-*O*-glucoside, and quercetin-3-*O*-D-galactopyranoside close by some critical iridoid glycosides like aucubin, catalpol, acetonide, and verbascoside have similarly been represented. (*E*)-β-farnesene, (*E*)-α-bergamotene, and sesquiterpenes (*E*)-β-caryophyllene are the most huge terpenoids itemized. The *Plantago* leaf extracts additional metallic components like arsenic, cadmium, copper, cobalt, iron, nickel, lead, zinc, magnesium, sodium, calcium, and phosphorus.

3.27.5 Pharmacological Properties

P. lanceolata herb contains 2–6.5% mucilage made up of at least four polysaccharides; 6.5% tannins; iridoid glycosides, including 0.3–2.5% aucubin and 0.3–1.1% catalpol; more than 1% silicic acid; phenolic carboxylic acid (protocatechuic acid); flavonoids (apigenin, luteolin); minerals, including essential zinc, potassium, silicic acid; and saponin.

Plantago is a safe and efficient way of dealing with bleeding, which immediately strengthens blood flow and promotes the mend of injured tissue. The leaves contain mucilage, tannins, and silicic acid. Their extract has antibacterial properties. They have an unpleasant taste and are astringent, demulcent, somewhat expectorant, hemostatic, and ophthalmic. Internally used for various complaints like diarrhoea, gastritis, peptic ulcer disease, irritable bowel syndrome, bleeding, cystitis, bronchitis, catarrh, sinusitis, asthma, and hay fever. They are employed exteriorly in the handling of skin inflammation, severe ulcers, cuts, itching, etc. Hot leaves are used as a wet cloth for wounds, swelling, etc. Roots are a remedy for rattlesnake stings, used in equal parts. Seeds are utilized to destroy worms. The plantain seeds contain up to 30% mucilage in the intestines, acting as a laxative in large quantities and a cooling lining. Sometimes the seed pods are employed with no seeds. Distilled water made from plants is an excellent eye ointment.

3.27.6 Scientific Investigation for the Management of Cancer

P. lanceolata L. is utilized in Iraqi old stories medication to treat wounds and its concentrate is recommended by certain botanists for disease patients.

3.27.7 Reported Mechanism of Action as Anti-cancer Activity

P. lanceolata leaf extricate specifically restrained the multiplication of CAL51 triple-negative bosom malignant growth cells and showed a minor impact on the other bosom disease cell types examined. Accordingly, *P. lanceolata* is a probable normal wellspring of the specific enemy of triple-negative breast cancer drugs.

3.27.8 Toxicological Aspects

At high doses, the herb induced cytopathic morphological changes. The clonogenic assay showed low colony formation in the exposed cells, especially CAL51 cells. Furthermore, the HPLC study revealed that the methanolic extract contained important flavonoid glycosides, especially rutin, myricetin quercetin, and kaempferol.

3.27.9 Conclusion and Future Aspects

The leaf extract is reliably non-toxic with strong hepato-protective and wound-healing activities; however, data about the responsible constituents is little and further research is required and needs to be further investigated.

3.28 *PODOPHYLLUM PELTATUM* OR *PODOPHYLLUM HEXANDRUM*

3.28.1 Background

Podophyllum is an herbaceous perennial plant in the family Berberidaceae, described as a genus by Linnaeus in 1753. *P. peltatum*, with common names may-apple, American mandrake, wild mandrake, and ground lemon, is prevalent athwart most of the eastern United States and south-eastern Canada.

Kingdom	**Plantae**
Division	Tracheophyta
Class	Magnoliopsida
Order	Ranunculales
Family	Berberidaceae
Genus	*Podophyllum*
Species	*Peltatum*
Binomial name	*Podophyllum peltatum* L.

3.28.2 Ethnomedical Considerations

P. peltatum or *P. hexandrum* is also the most fertile plant in historical history due to its long history of use by the Himalayan natives (Figure 3.28). This plant is conventionally applied to treat colds, constipation, peptic ulcers, inflammatory sensations, erysipelas, psychiatric disorders, disease, dynamic and inflammatory skin conditions, cancer of the brain, bladder, and lungs, venereal warts, monocytoid leukaemia, and disease of Hodgkin and non-Hodgkin's lymphoma.

FIGURE 3.28 Parts of *P. peltatum*.

3.28.3 Morphology and Pharmacognostic Characteristics

3.28.3.1 Morphology

Mayapples are wild plants, usually grown in colonies taken from a sole root. The stalks rise to 30–40 cm in height, with hand-shaped umbrella leaves up to 20–40 cm wide and three to nine shallow lobes for deep cuts. The plant produces multiple stems from the creeping rhizomes. Some stems have a single leaf and do not bear flowers or fruits, while the flowering stems produce two or more leaves with one to eight flowers in an axle stuck between the apical leaves. The blossoms are white, yellow, or red, 2–6 cm wide and with six to nine leaves, and ripen into a juicy green, yellow, or red fruit 2–5 cm long. All fractions of the plant are noxious, including raw fruit, but if the fruit is yellow, it can be securely consumed in small amounts with the seeds extracted. Ripe fruits do not produce toxins.

Podophyllum rhizomes come up in the form of sub-cylindrical fragments 5–20 cm long and 5–6 mm thick in the internode area and 15 mm in diameter. Fragments show branches from time to time. It demonstrates the marks of the aerial stems and the coming roots. The exterior is even or wrinkly and dark red. It shows a small but symbolic aroma and a bitter, sour taste. The rhizome breaks with a short, horned break. The opposite cut area shows starchy white circles with long veins.

3.28.3.2 Microscopic Characteristics

The opposite part of the *Podophyllum* rhizome represents the black epidermis and one or two deposits that are fabricated of lifeless cells. The outer cortex is fabricated of a parenchyma with thin walls and collenchymatous tissue, while the inner cortex contains a ring of small arteries. The central pith is parenchymatous with small stone cells. Certain parenchymal cells in the nodular region show crystals of calcium oxalate and most cells indicate the attendance of starch grains.

1. *Epiblema:* A single layer, yellowish-yellow, whose cells protrude slightly. The external and radial barriers are coated with suberin.
2. *Exodermis:* It has one covering right away below epiblema and has small cells with small walls.
3. *Cortex:* It has 18–22 layers of normal parenchymatous cells with wave walls and valleys with interlocking spaces. They are full of whole grain starch but without any crystals of calcium oxalate.
4. *Endodermis:* It has a single layer and forms the deepest layer of the left region. Tiny cells have very bright Casparian fibres.
5. *Vascular bonds:* Radial, alternate, groups varying from 4 to 9 and with exarch protoxylems.

3.28.4 Phytochemistry

American *Podophyllum* encloses 4–5% of podophyllum resin, while Indian varieties contain 7–16%. Indian *Podophyllum* resin provides a yield of 42–46% crystalline podophyllotoxin. Tannin was identified at the root of this plant by Wallis. The percentage of solid fibre is 10.5%, podophyllin is 11.5–15.56%, and podophyllotoxin is 3.19–4.10%. In addition to these compounds, small amounts of podophyllic acid, podophyllocersin, starch, tannins, pertatin, pertatin, and picropodophyllin are also available. The roots of this type are well known in pharmacology as a source of podophyllin resin. The resin produces podophyllotoxin as a major lignan and is produced by reduction of phenylpropanoid intermediates. The podophyllotoxin content is high (>5%) in *P. hexandrum* in addition to *P. peltatum* and that is why it gains more value. Roots of *P. peltatum* contain only 0.25% podophyllotoxin. *P. peltatum* contains 3.5–6% of active resin lignans, which are C18 biosynthetically compounds in flavonoids, and are obtained by the reduction of two C6–C3 units. The most important ingredient in podophyllum resin is podophyllotoxin (20% in American *Podophyllum*) and the highest value is about 40% in Indian *Podophyllum*. In addition, it contains α-peltatin (10%) and β-peltatin (5%). It is worth mentioning here that a large amount of lignan glycosides are also present in the plant, but due to its soluble properties in the water, they are almost eliminated during normal resin processing. Interestingly, all three of the chemical elements mentioned above exist within the Free State and as their glycosides. Indian *Podophyllum* does not contain α- and β-peltatin. The resin also contains dimethyl podophyllotoxin closely related to its glycoside and dehydro-podophyllotoxin and quercetin-A tetra-hydroxy flavonol.

3.28.5 Pharmacological Effects

The cancer-fighting activity of this plant is associated with the release of podophyllotoxin extracted from its root. It is very rare to fix podophyllotoxin biosynthetically, so its value is beyond doubt. Biotechnological extraction is available. This plant also has an antimitotic activity indicated by its extract called podophyllin. It is widely used in the treatment of ovarian cancer in women. Podophyllotoxin, podophyllin, and berberine are different types of lignin released from rhizome. *P. hexandrum* has anti-tumour properties because it can prevent mitosis at the metaphase level and prevent the proliferation of microtubules in the right way. Different types of cancer that can be treated effectively by this plant include lung cancer, testicular cancer, neuroblastoma, hepatoma, and other tumours.

Podophyllotoxin and D-glycoside lignins are released from the rhizome of this plant and have anti-tumour activity. Podophyllotoxin is said to be a medicinal compound and is a largely vital lignin. It has shown many cytotoxic activities. Some scientists corrected the podophyllotoxin results and realized that these substances were medically active in the event of cytotoxic activity at the micromolecular level. Uden et al. studied the phenylpropanoid and said that it is a derived lignan of podophyllotoxin.

3.28.6 Scientific Investigation for the Management of Cancer

Podophyllotoxin is a significant plant-inferred regular item and has subsidiaries, for example, etoposide and teniposide, which have been utilized as treatments for malignant growths and venereal mole. PTOX structure is firmly connected with the aryltetralin lactone lignans that have antineoplastic and antiviral exercises.

3.28.7 Reported Mechanism of Action as Anti-cancer Activity

Podophyllotoxin forestalls cell development through the polymerization of tubulin, prompting cell cycle capture and concealment of the arrangement of the mitotic-axles microtubules.

3.28.8 Toxicological Aspects

This plant is well known for its various beneficial uses, but there are also some side effects associated with its use. The side effects include bloody diarrhoea, harsh stomach ache, hallucinations, muscle paralysis, kidney failure, breathing failure, neuropathy, and encephalopathy. The employment of this plant is contraindicated in pregnancy.

3.28.9 Conclusion and Future Aspects

Several investigations have been performed for further preclinical studies that are warranted to explore the molecular mechanisms of these agents in treatment of cancer and their possible potential to overcome the chemoresistance of tumour cells.

3.29 *ADENANTHERA PAVONINA*

3.29.1 Background

A. pavonina is known by a number of common names, including red bead tree, and redwood. The tree is broadly developed in tropical areas as an ornament and is made naturally in many countries. The scientific name is derived from the combination of the two Greek words: "Aden," meaning gland, and "anthera" meaning anther. *A. pavonina* is a tree commonly called redwood and red bread; it is a leafy tree, 18–25 m tall, straight, and 60 cm wide.

Kingdom	**Plantae**
Division	Tracheophyta
Class	Magnoliopsida
Order	Fabales
Family	Fabaceae
Genus	*Adenanthera*
Species	*pavonina*
Binomial name	*Adenanthera pavonina* L.

3.29.2 Ethnomedical Considerations

A. pavonina, known as the red bead tree, has long been used in traditional medicine (Figure 3.29). Its various components are used to fight many types of diseases. This plant is often used for various ailments such as asthma, abscesses, diarrhoea, gout, inflammation, rheumatism, tumours, and ulcers, and as a tonic. It is reported to have various functions such as antioxidant and antibacterial action, anti-inflammatory,

FIGURE 3.29 Photograph of *Adenanthera pavonina*.

antimicrobial activity, antipyretic activity, hepatoprotective activity, hypolipidemic activity, hypolipidemic activity, wound-healing function, and antimalarial activity. Phytochemical research exposed the occurrence of glycosides, alkaloid tannins, and flavonoids.

3.29.3 Pharmacognostic Characteristics

3.29.3.1 Morphology

It is found in the Sub-Himalayan tract, ascending up to an altitude of 1,200 m in Assam, Maharashtra, Sikkim, West Bengal, Meghalaya, and Gujarat, and is regularly known as redwood. The main essential nutrients are flavonoid compounds. It is utilized as an antibacterial paste and is used to treat abscesses and inflammation. A medium-sized deciduous tree, *A. pavonine*, grows 6–15 m in height. It is usually stout, with dark brown to grey bark, and a spreading crown. The seeds are tightly tied, lens-shaped, bright red in colour, and stick to pods. The seed coat is smooth, shiny, bony, and very strong and usually has no cracks. The pods are leathery, curved, and twisted in deterioration to reveal 8–12 display seeds. The leaves are doubled. They are green at the top and blue at the bottom. They are yellow with age. The bark is dark brown or brown on the outside and has a whitish tinge on the inside. It is difficult for old trees with long cracks. A small, yellow flower grows on the tips of thick hanging flowers. They are small, creamy yellow in colour, and fragrant. Each flower is shaped like a five-leaf star. The wood is red in colour and extremely strong. It is durable and is used for construction purposes.

3.29.3.2 Macroscopic Characteristics

The opposite part of the bark shows normal microscopic features. Cork is an outer cover, containing several sheets of cork cells. They occur as normal lines of small cells with a thick, flat polygonal wall, packed close to the radial lines that die in maturity, and are produced in the cork cambium. In the periderm, phellogen have a persistent presence and are self-evident. Phloem is characterized by the collapse of cells. Scattered fragments of stone cells originate in the area of the cortex. The circular parenchyma cells in the cortex region divide into medullary radiation, which is like a tube in the inside fraction of the cortex. Medullary radiation tends to coalesce as it approaches the inner surface of a nearly parallel line, the phellem layer (cork) is directed outwards, while the phelloderm layer is directed inwards. Phellogen cells are rectangular and flat, appearing radial in the opposite phase. The phellem is made up of two or more layers of cells in a bright radial area but tangentially longer due to secondary growth. Phelloderm is two or more layers of rectangular cells. The periderm and cortex are separated by a layer of cells. The cortex region indicates the presence of simple crystals of calcium oxalate. Few mucilaginous cells exist and appear on their own. Phloem is characterized by the collapse of cells. Scattered fragments of stone cells are found in the area of the cortex. The circular parenchyma cells in the cortex region divide into medullary radiation, which is like a tube in the inner part of the cortex. The medullary rays tend to merge as they approach the inner stem as almost parallel lines.

3.29.4 Phytochemistry

It is believed to be rich in flavonoids, especially gallic acid, terpenoids, tannin, sterols (beta-sitosterol, beta-sitosterol-3β-D-glucoside), triterpenoids (nonacosane and hentriacontane), and saponins (sapogenins). Phytochemical studies indicate the presence of steroids, glycosides, and saponins in the offspring and pods. A new compound of the five-component lactone ring, Bonin, was separated from the soluble fraction of methanol. *A. Pavonina* seeds contain an active anti-inflammatory system, *O*-acetyl ethanolamine. The leaves contain octacosanol, dulcitol, beta sitosterol glucosides, and stigmasterol. The bark contains stigmasterol glucoside. The seeds contain HCN glucoside, lignoceric corrosive, dulcitol, stigmasterol, stigmasterol glucoside, and polysaccharide. The seeds are a promising source of galactomannans and polysaccharides.

A. pavonina bark was found to contain a reducing sugar (1.01%) as glucose. The percentage of different amino acids present in crude protein (5.25%) is aspartic acid (0.10%), threonine (0.24%), glutamic acid (0.52%), serine (0.08%), alanine (0.07%), glycine (0.09%), valine (0.10%), isoleucine (0.06%), methionine (0.13%), tyrosine (0.27%), lysine (0.88%), arginine (0.25%), and histidine (0.11%). The composition of fatty acids was found to be lauric (5.23%), stearic acid (8.93%), palmitic (38.16%), and oleic acid (6.29%).

3.29.5 Pharmacological Effects

A. pavonina is a remedial plant of traditional use that presents several scientific studies related to its biological activities. Based on their different derivatives, several activities were evaluated and confirmed by *in vitro* and *in vivo* studies by the various researchers as reported below.

3.29.5.1 Cancer-fighting Work

A decoction of *A. pavonina* L. and *Thespesia populnea* L. have been tested for cytotoxicity and anti-proliferation activity against HEp-2 cells using lactate dehydrogenase release, (3-(4,5-dimethylthiazol-2-yl)-2,5-diphenyltetrazolium bromide), and sulforhodamine B. The presentation of apoptosis is characterized by fluorescence microscopy contaminated with an ethidium bromide/acridine dye orange blend. Additionally, the salty aqueous shrimp lethality assay showed a level of EC50 at high concentrations (1.96 mg/ml).

3.29.5.2 Antioxidant Activity

The scavenging activity of the methanolic extract of the leaves of *A. pavonina* Linn. was evaluated to find the ability to counteract oxidative damage. Scavenging activity was evaluated by DPPH free radical plus nitric oxide anion scavenging assays amid ascorbic acid as standard. Antioxidant activity of methanolic extracts of leaf and bark using DPPH scavenging activity comparison shows that leaf extract has a little higher antioxidant activity than the bark extract. The leafy part is more active in respect of its antioxidant activity than the bark, though the leaf part is lacking flavonoids.

3.29.6 Scientific Investigation for the Management of Cancer

A decoction arranged in the mix with these two plants, *A. pavonina* L. has been utilized for quite a long time to fix malignant growth in patients.

3.29.7 Reported Mechanism of Action as Anti-cancer Activity

Lindamulage and Soysa reported on the traditional use of *A. pavonina* derivatives in cancer therapy. Sowemimo et al. (2009) showed no cytotoxicity of the ethanolic fruit extract of *A. pavonina* on HeLa cell lines. Ferreira et al. (2011) evaluated the cytotoxic potential of an ethanol extract (50 μg/ml) from *A. pavonina* seeds in cancer cell lines. Araujo et al. (2019) evaluated the effect of *A. pavonina*'s antiproliferative activity on cancer cells. Pannonia seed powder is enzymatically treated with amylase, cellulase, and protease.

3.29.8 Toxicological Aspects

Apoptotic cells at higher concentrations suggest that the effect of the decoction is cytocidal, at least at higher concentrations.

3.29.9 Conclusion and Future Aspects

The study suggests that the antioxidant activity of these extracts might be helpful in preventing or slowing the progress of oxidative stress-related diseases like cancer. Identification of the antioxidant constituents of the plant which are helpful for the anti-cancer properties is yet to be studied.

BIBLIOGRAPHY

Abate, L., Abebe, A., & Mekonnen, A. (2017). Studies on antioxidant and antibacterial activities of crude extracts of *Plantago lanceolata* leaves. *Chemistry International, 3*, 277–287.

Abdelhalim, A., Aburjai, T., Hanrahan, J., & Abdel-Halim, H. (2017). Medicinal plants used by traditional healers in Jordan, the Tafila region. *Pharmacognosy Magazine, 13*, S95.

Abdel-Magied, E. M., Abdel-Rahman, H. A., & Harraz, F. M. (2001). The effect of aqueous extracts of *Cynomorium coccineum* and *Withania somnifera* on testicular development in immature Wistar rats. *Journal of Ethnopharmacology, 75*(1), 1–4.

Aburjai, T., Hudaib, M., Tayyem, R., Yousef, M., & Qishawi, M. (2007). Ethnopharmacological survey of medicinal herbs in Jordan, the Ajloun Heights region. *Journal of Ethnopharmacology, 110*, 294–304.

Adedapo, A. A., Abatan, M. O., Idowu, S. O., & Olorunsogo, O. O. (2005). Toxic effects of chromatographic fractions of *Phyllanthus amarus* on the serum biochemistry of rats. *Phytotherapy Research, 19*, 812–815.

Adedapo, A. D., Osude, Y. O., Adedapo, A. A., Moody, J. O., Adeagbo, A. S., Olajide, O. A., & Makinde, J. M. (2009). Blood pressure lowering effect of *Adenanthera pavonina* seed extract on normotensive rats. *Records of Natural Products, 3*(2), 82.

AdnanTuama, A., & Ahmed Mohammed, A. (2019). Phytochemical screening and *in vitro* antibacterial and anticancer activities of the aqueous extract of *Cucumis sativu*. *Saudi Journal of Biological Sciences, 26*(3), 600–604.

Ahmad, M. S., Bano, S., & Anwar, S. (2015). Cancer ameliorating potential of *Phyllanthus amarus*: *In vivo* and *in vitro* studies against aflatoxin B1 toxicity. *Egyptian Journal of Medical Human Genetics, 16*, 343–353.

Ahmad, W., Hasan, A., Ansari, A., & Tarannum, T. (2010). *Curcuma longa* Linn.: A review. *Hippocratic Journal of Unani Medicine, 5*, 179–190.

Ahmad, M. K., Mahdi, A. A., Shukla, K. K., Islam, N., Rajender, S., Madhukar, D., Shankhwar, S. N., & Ahmad, S. (2010). *Withania somnifera* improves semen quality by regulating reproductive hormone levels and oxidative stress in seminal plasma of infertile males. *Fertility and Sterility, 94*(3), 989–996.

Ahmed, S., Khatri, M. S., & Hasan, M. M. (2017). Plants of family Lamiaceae: A promising hand for new antiurolithiatic drug development. *World Journal of Pharmacy and Pharmaceutical Sciences, 6*(7), 90–96.

Akram, M., Uddin, S., Ahmed, A., Usmanghani, K., Abdul Hannan, A., Mohiuddin, E., & Asi, M. (2010). *Curcuma Longa* and curcumin: A review. *Romanian Journal of Biology: Plant Biology, 55*, 65–70.

Alagawany, M., Ashour, E. A., & Reda, F. M. (2016). Effect of dietary supplementation of garlic (*Allium sativum*) and turmeric (*Curcuma longa*) on growth performance, carcass traits, blood profile and oxidative status in growing rabbits. *Annals of Animal Science, 16*, 489–505.

Al-Hwaiti, M. S., Alsbou, E. M., Abu Sheikha, G., Bakchiche, B., Pham, T. H., Thomas, R. H., & Bardawee, S. K. (2021). Evaluation of the anticancer activity and fatty acids composition of "Handal" (*Citrullus colocynthis* L.) seed oil, a desert plant from South Jordan. *Food Science & Nutrition, 9*(1), 282–289.

Allegra, A., Tonacci, A., Pioggia, G., Musolino, C., & Gangemi, S. (2020). Anticancer activity of *Rosmarinus officinalis* L.: Mechanisms of action and therapeutic potentials. *Nutrients*, *12*(6), 1739.

Alsaraf, K. M., Mohammad, M. H., Al-Shammari, A. M., & Abbas, I. S. (2019). Selective cytotoxic effect of *Plantago lanceolata* L. against breast cancer cells. *Journal of the Egyptian National Cancer Institute*, *31*, 10.

Al-Saraireh, Y. M., Youssef, A. M. M., Za'al Alsarayreh, A., Al Hujran, T. A., Al-Sarayreh, S., Al-Shuneigat, J. M., & Alrawashdeh, H. M. (2021). Phytochemical and anti-cancer properties of *Euphorbia hierosolymitana* Boiss. crude extracts. *Journal of Pharmacy & Pharmacognosy Research*, *9*(1), 13–23.

Alzoubi, K. H., Khabour, O. F., Alkofahi, A. S., & Mhaidat, N. M. (2021). Anticancer and antimutagenic activity of *Silybum marianum* L. and *Eucalyptus camaldulensis* Dehnh. against skin cancer induced by DMBA: *In vitro* and *in vivo* models. *Pakistan Journal of Pharmaceutical Sciences*, *34*(3), 987–993.

Amalraj, A., Varma, K., Jacob, J., Divya, C., Kunnumakkara, A. B., Stohs, S. J., & Gopi, S. (2017). A novel highly bioavailable curcumin formulation improves symptoms and diagnostic indicators in rheumatoid arthritis patients: A randomized, double-blind, placebo-controlled, two-dose, three-arm and parallel-group study. *Journal of Medicinal Food*, *20*, 1022–1030.

Ara, A., Arifuzzaman, M., Ghosh, C. K., Hashem, M. A., Ahmad, M. U., Bachar, S. C., & Sarker, S. D. (2010). Anti-inflammatory activity of *Adenanthera pavonina* L., Fabaceae, in experimental animals. *Revista Brasileira de Farmacognosia*, *20*, 929–932.

Ardalani, H., Avan, A., & Ghayour-Mobarhan, M. (2017). Podophyllotoxin: A novel potential natural anticancer agent. *Avicenna Journal of Phytomedicine*, *7*(4), 285–294.

Baquar, S. R. (1995). The role of traditional medicine in rural environment. In S. Issaq (Ed.), *Traditional medicine in Africa* (pp. 41–142). East Africa Educational Publishers Ltd.

Barth, A., Müller, D., & Dürrling, K. (2002). *In vitro* investigation of a standardized dried extract of *Citrullus colocynthis* on liver toxicity in adult rats. *Experimental and Toxicologic Pathology*, *54*, 223–230.

Bauer, B., Kavrakovski, Z., & Kostik, V. (2013) *An ethno-pharmacological and toxicological review of Ecballium elaterum* (L.) A. Rich. International Conference on Natural Products Utilization from Plants to Pharmacy Shelf (ICNPU), 3–6 November 2013.

Begum, A., Sandhya, S., Vinod, K. R., Reddy, S., & Banji, D. (2013). An in-depth review on the medicinal flora *Rosmarinus officinalis* (Lamiaceae). *Acta scientiarum polonorum. Technologia alimentaria*, *12*(1), 61–74.

Bhandari, P. R. (2015). *Crocus sativus* L. (saffron) for cancer chemoprevention: A mini review. *Journal of Traditional and Complementary Medicine*, *5*(2), 81–87.

Burge, M. (1980). The use of lemon balm (*Melissa officinalis*) for attracting honeybee swarms. *Bee World*, *61*(2), 44–46.

Chaemsawang, W., Prasongchean, W., Papadopoulos, K. I., Ritthidej, G., Sukrong, S., & Wattanaarsakit, P. (2019). The effect of okra (*Abelmoschus esculentus* (L.) Moench.) seed extract on human cancer cell lines delivered in its native form and loaded in polymeric micelles. *International Journal of Biomaterials*, *2019*, 9404383.

Chaurasia, O. P., Ballabh, B., Tayade, A., Kumar, R., Kumar, G. P., & Singh, S. B. (2012).*Podophyllum* L.: An endergered and anticancerous medicinal plant—An overview. *Indian Journal of Traditional Knowledge*, *11*(2), 234–241.

Danin, A., & Künne, I. (1996). *Origanum jordanicum* (Labiatae), a new species from Jordan, and notes on the other species of *O. sect.* Campanulaticalyx. *Willdenowia*, 601–611.

De Sousa, A. C., Gattass, C. R., Alviano, D. S., Alviano, C. S., Blank, A. F., & Alves, P. B. (2004). *Melissa officinalis* L. essential oil: Antitumoral and antioxidant activities. *Journal of Pharmacy and Pharmacology*, *56*(5), 677–681.

Desai, V. R., Kamat, J. P., & Sainis, K. B. (2002). An immunomodulator from *Tinospora cordifolia* with antioxidant activity in cell-free systems. *Journal of Chemical Sciences*, *114*(6), 713–719.

Dietz, M., Machill, S., Hoffmann, H. C., & Schmidtke, K. (2013). Inhibitory effects of *Plantago lanceolata* L. on soil *N* mineralization. *Plant and Soil*, *368*(1), 445–458.

Dikshit, V., Damre, A. S., Kulkarni, K. R., Gokhale, A., & Saraf, M. N. (2000). Preliminary screening of imunocin for immunomodulatory activity. *Indian Journal of Pharmaceutical Sciences*, *62*, 257.

Diwanay, S., Chitre, D., & Patwardhan, B. (2004). Immunoprotection by botanical drug on experimental metastasis. *Journal of Ethanopharmacology*, *90*, 223–237.

Du, H., Wang, J., Hu, Z., & Yao, X. (2008). Quantitative structure-retention relationship study of the constituents of saffron aroma in SPME-GC-MS based on the projection pursuit regression method. *Talanta*, *77*, 360–365.

Dubey, S., Singh, M., Nelson, A., & Karan, D. (2021). A perspective on *Withania somnifera* modulating antitumor immunity in targeting prostate cancer. *Journal of Immunology Research*, 2021, 9483433.

Durna Daştan, S., Daştan, T., Çetinkaya, S., Ateşşahin, D., & Karan, T. (2016). Evaluation of *in vitro* anticancer effect of *Plantago major* L. and *Plantago lanceolata* L. leaf extracts from Sivas. *Cumhuriyet Üniversitesi Sağlık Bilimleri Enstitüsü Dergisi*, 1(1), 7–14.

El-Gengaihi, S., El-Hamid, A., Sr., & Kamel, A. (2009). Anti-inflammatory effect of some cucurbitaceous plants. *Herba Polonica*, *55*(4), 119–126.

El-Ghorab, A. H., Behery, F. A., Abdelgawad, M. A., Alsohaimi, I. H., Musa, A., Mostafa, E. M., Altaleb, H. A., Althobaiti, I. O., Hamza, M., Elkomy, M. H., Hamed, A. A., Sayed, A. M., Hassan, H. M., & Aboseada, M. A. (2022). LC/MS profiling and gold nanoparticle formulation of major metabolites from *Origanum majorana* as antibacterial and antioxidant potentialities. *Plants*, *11*, 1871.

Erenler, R., Telci, I., Elmastaş, M., Akşit, H., Gül, F., Tüfekçi, A. R., Demirtaş, I., & Kayir, Ö (2018). Quantification of flavonoids isolated from *Mentha spicata* in selected clones of Turkish mint landraces. *Turkish Journal of Chemistry*, *42*, 1695–1705.

Escribano, J., Alonso, G. L., Coca-Prados, M., & Fernandez, J. A. (1996). Crocin, safranal and picrocrocin from saffron (*Crocus sativus* L.) inhibit the growth of human cancer cells *in vitro*. *Cancer Letters*, *100*, 23–30.

Faraji, P., Araj-Khodaei, M., Ghaffari, M., & Ezzati Nazhad Dolatabadi, J. (2022). Anticancer effects of *Melissa officinalis*: A traditional medicine. *Pharmaceutical Sciences Pharmaceutical Sciences*, *28*(3), 355–364.

Fayera, S., Babu, G. N., Dekebo, A., & Bogale, Y. (2018). Phytochemical investigation and antimicrobial study of leaf extract of *Plantago lanceolata*. *Natural Products Chemistry and Research*, *6*(2), 1–8.

Foo, L. Y., & Wong, H. (1992). Phyllanthusiin D, an unusual hydrosable tannin from *Phyllanthus amarus*. *Phytochemistry*, *31*, 711–713.

Giri, A., & Lakshmi Narasu, M. (2000). Production of podophyllotoxin from *Podophyllum hexandrum*: A potential natural product for clinically useful anticancer drugs. *Cytotechnology*, *34*(1), 17–26.

Gomez-Flores, R., Calderon, C. L., Scheibel, L. W., Tamez-Guerra, P., Rodriguez-Padilla, C., Tamez-Guerra, R., & Weber, R. J. (2000). Immunoenhancing properties of *Plantago major* leaf extract. *Phytotherapy Research*, *14*(8), 617–622.

González-Minero, F. J., Bravo-Díaz, L., & Ayala-Gómez, A. (2020). *Rosmarinus officinalis* L. (rosemary): An ancient plant with uses in personal healthcare and cosmetics. *Cosmetics*, *7*(4), 77.

Hameed, I. H., & Mohammed, G. J. (2017). *Phytochemistry, antioxidant, antibacterial activity, and medicinal uses of aromatic (medicinal plant Rosmarinus officinalis): Aromatic and medicinal plants* (pp. 175–189). Intech Open.

Hegazi, G. A., & El-Lamey, T. M. (2011). *In vitro* production of some phenolic compounds from *Ephedra alata* Decne. *Journal of Applied Environmental and Biological Sciences*, *1*(8), 158–163.

Hosseinzadeh, H., & Ghenaati, J. (2006). Evaluation of the antitussive effect of stigma and petals of saffron (*Crocus sativus*) and its components, safranal and crocin in guinea pigs. *Fitoterapia*, *77*, 446–448.

Ilango, K. B., Rajkumar, P., Vetrivel, D., Brinda, P., & Mishra, M. (2011). Safety evaluation of *Abelmoschus esculentus* polysaccharide. *International Journal of Pharmaceutical Sciences Review and Research*, *10*, 106–110.

Itoro, E., Ukana, D., & Ekaete, D. (2013). Phytochemical screening and nutrient analysis of *Phyllanthus amarus*. *Asian Journal of Plant Science and Research*, *3*, 116–122.

Jafargholizadeh, N., Zargar, S. J., & Aftab, Y. (2018). The cucurbitacins D, E, and I from *Ecballium elaterium* (L.) upregulate the *LC3* gene and induce cell-cycle arrest in human gastric cancer cell line AGS. *Iranian Journal of Basic Medical Sciences*, *21*(3), 253–259.

Jagetia, G. C. (2019). Anticancer activity of Giloe, *Tinospora cordifoila* (Willd.) Miers Ex Hook F & Thoms. *International Journal of Complementary and Alternative Medicine*, *12*(2), 79–85.

Jain, N. (2012). A review on *Abelmoschus esculentus*. *Pharmacacia*, *1*, 1–8.

Jain, S. K., & Khurdiya, D. S. (2004). Vitamin c enrichment of fruit juice based ready-to-serve beverages through blending of Indian gooseberry (*Emblica officinalis* Gaertn.) juice. *Plant Foods for Human Nutrition*, *59*(2), 63–66.

Jaradat, N., Hussen, F., & Al-Ali, A. (2015). Preliminary phytochemical screening, quantitative estimation of total flavonoids, total phenols and antioxidant activity of *Ephedra alata* Decne. *Journal of Materials and Environmental Science*, *6*(6), 1771–1778.

Jarret, R. L., Wang, M. L., & Levy, I. J. (2011). Seed oil and fatty acid content in okra (*Abelmoschus esculentus*) and related species. *Journal of Agricultural Food Chemistry*, *59*(8), 4019–24.

Jena, P. K., Dinda, S. C., & Ellaiah, P. (2011). Phytochemical investigation and simultaneous study on antipyretic, anticonvulsant activity of different leafy extracts of *Smilax zeylanica* Linn. *Oriental Pharmacy and Experimental Medicine*, *12*, 123–127.

Jena, P. K., Nayak, B. S., Dinda, S. C., & Ellaiah, P. (2011). Investigation on phytochemicals, anthelmintic and analgesic activities of *Smilax zeylanica* Linn. leafy extracts. *Asian Journal of Chemistry*, *23*(10), 4307–4310.

Joseph, B., & Raj, S. J. (2011). An overview: Pharmacognostic property of *Phyllanthus amarus* Linn. *International Journal of Pharmacology*, *7*, 40–45.

Jurič, A., Huđek Turković, A., Brčić Karačonji, I., Prđun, S., Bubalo, D., & Durgo, K. (2022). Cytotoxic activity of strawberry tree (*Arbutus Unedo* L.) honey, its extract, and homogentisic acid on CAL 27, HepG2, and Caco-2 cell lines. *Arhiv za higijenu rada i toksikologiju*, *73*(2), 158–168.

Kalam, M. A., Malik, A. H., Ganie, A. H., & Butt, T. A. (2021). Medicinal importance of papra (*Podophyllum hexandrum* Royle.) in Unani system of medicine. *Journal of Complementary and Integrative Medicine*, *18*(3), 485–490.

Khan, F. A., Zahoor, M., Ullah, N., Khan, S., Khurram, M., Khan, S., & Ali, J. (2014). A general introduction to medicinal plants and *Silybum marianum*. *Life Science Journal*, *11*(9), 471–481.

Kirtikar, K. R., & Basu, B. D. (2006). *Indian medicinal plants* (Vol. 2,pp. 1417–1418). International Book Distributors.

Krishnaiah, D., Devi, T., Bono, A., & Sarbatly, R. (2009). Studies on phytochemical constituents of six Malaysian medicinal plants. *Journal of Medicinal Plant Research*, *3*(2), 67–72.

Krochmal, A. (1974). *Mayapple (Podophyllum peltatum L.)*, Research Paper No. 296 (Forest Service, US Department of Agriculture, Northeastern Forest Experiment Station).

Kumar, A., & Chauhan, G. S. (2010). Extraction and characterization of pectin from apple pomace and its evaluation as lipase (steapsin) inhibitor. *Carbohydrate Polymers, 82*, 454–459.

Lee, K., Kim, Y., Kim, D., Lee, H., & Lee, C. (2003). Major phenolics in apple and their contribution to the total antioxidant capacity. *Journal of Agricultural and Food Chemistry, 51*, 6516–6520.

Lindamulage, I. K. S., & Soysa, P. (2016). Evaluation of anticancer properties of a decoction containing *Adenanthera pavonina* L. and *Thespesia populnea* L. *BMC Complementary Medicine and Therapies, 16*, 70.

Mahendran, G., Kumar, S., & Rahman, V. L.-U. (2021). The traditional uses, phytochemistry and pharmacology of spearmint (*Mentha spicata* L.): A review. *Journal of Ethnopharmacology, 278*, 114266.

Mandlik Ingawale, D. S., & Namdeo, A. G. (2021). Pharmacological evaluation of ashwagandha highlighting its healthcare claims, safety, and toxicity aspects. *Journal of Dietary Supplements, 18*(2), 183–226.

Mantle, D., Lennard, T. W., & Pickering, A. T. (2000). Therapeutic applications of medicinal plants in the treatment of breast cancer: A review of their pharmacology, efficacy and tolerability. *Adverse Drug Reactions and Toxicological Reviews, 19*(3), 223–240.

Marzouk, B., Marzouk, Z., & Decor, R. (2009). Antibacterial and anticandidal screening of Tunisian *Citrullus colocynthis* Schrad. from medenine. *Journal of Ethnopharmacology, 125*, 344–349.

Mazid, M., Khan, T. A., & Mohammad, F. (2012). Medicinal plants of rural India: A review of use of medicinal plant by Indian folks. *Indo Global Journal of Pharmaceutical Sciences, 2*(3), 286–304.

Mehta, V., Chander, H., & Munshi, A. (2021). Mechanisms of anti-tumor activity of *Withania somnifera* (ashwagandha). *Nutrition and Cancer, 73*(6), 914–926.

Melese, E., Asres, K., Asad, M., & Engidawork, E. (2011). Evaluation of the antipeptic ulcer activity of the leaf extract of *Plantago lanceolata* L. in rodents. *Phytotherapy Research, 25*(8), 1174–1180.

Menegazzi, M., Masiello, P., & Novelli, M. (2020). Anti-tumor activity of *Hypericum perforatum* L. and hyperforin through modulation of inflammatory signaling, ROS generation and proton dynamics. *Antioxidants (Basel), 10*(1), 18.

Menyiy, N. E., et al. (2011). Medicinal uses, phytochemistry, pharmacology, and toxicology of *Mentha spicata. Journal of Medicinal Plants Research, 5*(20), 5142, 5147.

Mujahid, M., Ansari, V. A., Sirbaiya, A. K., Kumar, R., & Usmani, A. (2016). An insight of pharmacognostic and phytopharmacology study of *Adenanthera pavonina. Journal of Chemical and Pharmaceutical Research, 8*(2), 586–596.

Murali, A., Ashok, P., & Madhavan, V. (2010). Antioxidant effect of roots and rhizomes of *Smilax zeylanica* L.: An *in vivo* study. *Journal of Pharmacy Technology, 2*, 862–875.

Muruganantham, N., Solomon, S., & Senthamilselvi, M. M. (2016). Anti-cancer activity of *Cucumis sativus* (Cucumber) flowers against human liver cancer. *International Journal of Pharmaceutical and Clinical Research, 8*(1), 39–41.

Nmila, R., Gross, R., & Rchid, H. (2000). Insulinotropic effect of *Citrullus colocynthis* fruit extracts. *Planta Medica, 66*, 418–423.

Olajide, O. A., Echianu, C. A., Adedapo, A. D., & Makinde, J. M. (2004). Anti-inflammatory studies on *Adenanthera pavonina* seed extract. *Inflammopharmacology, 12*(2), 196–201.

Oluwatuyi, M., Kaatz, G. W., & Gibbons, S. (2004). Antibacterial and resistance modifying activity of *Rosmarinus officinalis. Phytochemistry, 65*(24), 3249–3254.

Palliyaguru, D. L., Singh, S. V., & Kensler, T. W. (2016). *Withania somnifera*: From prevention to treatment of cancer. *Molecular Nutrition & Food Research*, *60*(6), 1342–1353.

Panda, H. (1999). *Herbs cultivation and medicinal uses* (pp. 240–241). National Institute of Industrial Research.

Partha, G., & Rahaman, C. H. (2015). Pharmacognostic, phytochemical and anti-oxidant studies of *Adenanthera pavonina* L. *International Journal of Pharmacognosy and Phytochemical Research*, *7*(1), 30–37.

Patel, C. J., Tyagi, S., Kumar, U., Patel, S., Patel, P., & Bharat, C. (2013). Clinical benefits of milk thistle (*Silybum marianum*): A recent review. *Journal of Drug Discovery and Therapeutics*, *1*(1), 08–11.

Patocka, J., Bhardwaj, K., Klimova, B., Nepovimova, E., Wu, Q., Landi, M., Kuca, K., Valis, M., & Wu, W. (2020). *Malus domestica*: A review on nutritional features, chemical composition, traditional and medicinal value. *Plants (Basel)*, *9*(11), 1408.

Paul, S., Chakraborty, S., Anand, U., Dey, S., Nandy, S., Ghorai, M., Chatterjee Saha, S., Patil, M. T., Kandimalla, R., Proćków, J., & Dey, A. (2021). *Withania somnifera* (L.) Dunal. (Ashwagandha): A comprehensive review on ethnopharmacology, pharmacotherapeutics, biomedicinal and toxicological aspects. *Biomedicine & Pharmacotherapy*, *143*, 112175.

Perveen, S., Ashfaq, H., Ambreen, S., Ashfaq, I., Kanwal, Z., & Tayyeb, A. (2021). Methanolic extract of *Citrullus colocynthis* suppresses growth and proliferation of breast cancer cells through regulation of cell cycle. *Saudi Journal of Biological Sciences*, *28*(1), 879–886.

Podsędek, A., Wilska-Jeszka, J., Anders, B., & Markowski, J. (2000). Compositional characterisation of some apple varieties. *European Food Research and Technology*, *210*, 268–272.

Rai, M., Jogee, P. S., Agarkar, G., & Santos, C. A. (2016). Anticancer activities of *Withania somnifera*: Current research, formulations, and future perspectives. *Pharmaceutical Biology*, *54*(2), 189–197.

Rajendran, T. P. (2016) Biological resource base for traditional medicines. *Health, Nature and Quality of Life*, 41.

Rawal, S., Singh, P., Gupta, A., & Mohanty, S. (2014). Dietary intake of *Curcuma longa* and *Emblica officinalis* increases life span in *Drosophila melanogaster. BioMed Research International*, *2014*, 7.

Renilda Sophy, A. J., & Fleming, A. T. (2013). Evaluation of *Adenanthera pavonina* bark extracts for antioxidant activity and cytotoxicity against cancer cell lines. *International Journal of Science and Research*, *6*, 14.

Ribeiro-Santos, R., Carvalho-Costa, D., Cavaleiro, C., Costa, H. S., Albuquerque, T. G., Castilho, M. C., & Sanches-Silva, A. (2015). A novel insight on an ancient aromatic plant: The rosemary (*Rosmarinus officinalis* L.). *Trends in Food Science & Technology*, *45*(2), 355–368.

Romitha Lobo, A., Satish, S., & Shabaraya, A. R.. (2018). Review on pharmacological activities of *Malus domestica*. *International Journal of Pharma and Chemical Research*, *4*(4), 231–237.

Saggam, A., Tillu, G., Dixit, S., Chavan, P., Borse, G. S., Joshi, K., & Patwardhan, B. (2020). *Withania somnifera* (L.) Dunal.: A potential therapeutic adjuvant in cancer. *Journal of Ethnopharmacology*, *255*, 112759.

Saglam, C., Atakisi, I., Turhan, H., Kaba, S., Arslanoglu, F., & Onemli, F. (2004). Effect of propagation method, plant density, and age on lemon balm (*Melissa officinalis*) herb and oil yield. *New Zealand Journal of Crop and Horticultural Science*, *32*(4), 419–423.

Sayed, Z. I., & Badr, W. H. (2012). Cucurbitacin glucosides and biological activities of the ethyl acetate fraction from ethanolic extract of Egyptian *Ecballium elaterium*. *Journal of Applied Sciences Research*, *8*(2), 1252–1258.

Şengün, D. N., Karaca, İnciR., Saraç, N., Uğur, A., Fırat, A., Kaymaz, F. F., & Öztürk, H. S. (2021). Evaluation of the chemopreventive effects of *Hypericum perforatum* L. on DMBA-applied rat oral mucosa. *Archives of Oral Biology, 127,* 105139.

Shayista, C., Zahoor, A. K., & Phalestine, S. (2013). Medicinal importance of genus *Origanum*: A review. *Journal of Pharmacognosy and Phytotherapy, 5*(10), 170–177.

Shrivastava, J., Lambart, J., & Vietmeyer, N. (1996). *Medicinal plants: An expanding role in development.* World Bank Technical Paper, 320 p.

Sioud, F., Amor, S., Toumia, I., Lahmar, A., Aires, V., Chekir Ghedira, L., & Delmas, D. (2020). A new highlight of *Ephedra alata* Decne. properties as potential adjuvant in combination with cisplatin to induce cell death of 4T1 breast cancer cells *in vitro* and *in vivo. Cells, 9*(2), 362.

Soliman, M. S. M., Abdella, A., Khidr, Y. A., Hassan, G. O. O., Al-Saman, M. A., & Elsanhoty, R. M. (2021). Pharmacological activities and characterization of phenolic and flavonoid compounds in methanolic extract of *Euphorbia cuneata* Vahl. aerial parts. *Molecules, 26*(23), 7345.

Spengler, R. N. (2019). Origins of the apple: The role of megafaunal mutualism in the domestication of malus and rosaceous trees. *Frontiers in Plant Science, 10,* 617.

Tang, Y.-Q., Jaganath, I. B., Manikam, R., & Devi Sekaran, S. (2013). *Phyllanthus* suppresses prostate cancer cell, PC-3, proliferation and induces apoptosis through multiple signalling pathways (MAPKs, PI3K/Akt, NFB, and hypoxia). *Evidence-Based Complementary and Alternative Medicine,* 2013, 1–12.

Touihri, I., Kallech-Ziri, O., Boulila, A., Fatnassi, S., Marrakchi, N., Luis, J., & Hanchi, B. (2019). *Ecballium elaterium* (L.) A. Rich. seed oil: Chemical composition and antiproliferative effect on human colonic adenocarcinoma and fibrosarcoma cancer cell lines. *Arabian Journal of Chemistry, 12,* 2347–2355.

Ulbricht, C., Brendler, T., Gruenwald, J., Kligler, B., Keifer, D., Abrams, T. R., & Lafferty, H. J. (2005). Lemon balm (*Melissa officinalis* L.): An evidence-based systematic review by the natural standard research collaboration. *Journal of Herbal Pharmacotherapy, 5*(4), 71–114.

Verma, P. P. S., Singh, A., ur-Rehman, L., & Bahl, J. R. (2015) Lemon balm (*Melissa officinalis* L.) an herbal medicinal plant with broad therapeutic uses and cultivation practices: A review. *International Journal of Recent Advances in Multidisciplinary Research, 2*(11), 928–933.

Yoshidaand, Y., & Niki, E. (2003). Antioxidant effects of phytosterol and its components. *Journal of Nutritional Science and Vitaminology (Tokyo), 49*(4), 277–280.

Zhao, T., Sun, Q., Marques, M., & Witcher, M. (2015). Anticancer properties of *Phyllanthus emblica* (Indian gooseberry). *Oxidative Medicine and Cellular Longevity, 2015,* 950890.

4 Method of Extraction/ Isolation of the Phytoconstituents Responsible for Anti-cancer Activity

Ankur Joshi, Sapna Malviya, Neelesh Malviya, and Priyanka Soni

CONTENTS

DOI: 10.1201/9781003251712-4

4.1 INTRODUCTION

Cancer is one of the most dreadful diseases of the twenty-first century, with approximately 6 million cases reported each year. Lung, colon, prostate, and breast cancers are the second leading causes of death after cardiovascular disease. Mother Nature has bestowed upon us a vast treasure trove of valuable medicinal flora capable of curing a wide range of diseases. The use of characteristic herbs as a panacea for various maladies can be traced back to 1500 BC. The significant difference between Western and Ayurvedic cancer treatment is that Ayurveda does not include any harsh treatments.

Ayurvedic treatments were found to be more effective in treating chronic illnesses that were previously unsuitable for Western therapeutic practices. The Ayurvedic system of treatment is based on natural principles, and its components are based on extensive research into human physiology. Ayurveda defines good health as the perfect condition of an individual's physical, mental, social, and profound segments.

The goal of Ayurvedic treatment is to empower the body's self-healing capacities without any side effects by utilizing medicinal herbs to cleanse and bolster body tissues for recuperation. The first step is to bring the tridosha and triguna under control. Herbal medicines have been used successfully for hundreds of years in India, and they have been consumed globally in recent decades. The pharmaceutical market was worth $550 billion in 2004, and it is expected to be worth $900 billion by 2009. The current demand for medicinal plants is $14 billion per year, with a projected increase to $5 trillion by 2050. People in both developing and developed countries began using herbal drugs after learning about the dangerous side effects of synthetic drugs and their high costs.

Traditional Indian drugs have attracted professionals and specialists for their potential to cure cancer as a research-based foundation for hundreds of years. It has been estimated that 60% of the widely used (approved) drugs for cancer are derived from common sources. Herbal extracts have a high potential for disease cure and are commonly used in Ayurveda. The advantage of home-prepared decoctions is that they work in a systematic manner by supporting various organs of the body. The Ayurvedic classification of neoplasms is based on different clinical manifestations in relation to the tridoshas. Group I diseases include arbuda and granthi, for example, mamsarbuda (melanoma) and raktarbuda (leukaemia), mukharbuda (oral tumour), and so on. Group II diseases include those that can be classified as tumours, such as serious ulcers like tridosaj gulmas (tumours of the stomach like carcinomas of the liver or lymphomas and stomach). Visarpa (erysipelas), asadhya kamala (hopeless jaundice), and nadi vrana are examples of diseases in Group III (sinusitis).

AYUSH is an abbreviation for the medical systems practised in India, including Ayurveda, yoga and naturopathy, Unani, Siddha, and homoeopathy. These systems are based on specific medical philosophies and represent a way of healthy living with established concepts on disease prevention and health promotion. All of these systems take a holistic approach to health, disease, and treatment. As a result, there is renewed interest in AYUSH systems.

Charaka and *Sushruta Samhita* are two well-known Ayurvedic works that depict cancer as either fiery or non-provocative inflammations and classify them as granthi (minor neoplasm) or arbuda (major neoplasm). Ayurvedic writing distinguishes three body control frameworks: the sensory system (vata or air), the venous structure (pitta or fire), and the blood vessel structure (kapha or water), which work together to play out the body's typical capacity. *Sushruta* describes six different types of tumours. The first four tumours are benign and can be successfully treated in their early stages with the right treatments: paitika granthi (pitta dosha), medas granthi (affected fat tissue), vatika granthi (Vata dosha), and kaphaja granthi (kapha dosha). The last two tumours are malignant: rakta arbuda (blood tumour) and mamsa arbuda (muscle tumour). Cells divide and die normally, and the dead cells are replaced by new healthy cells. If dead cells are not replaced, they can form tumours, which are abnormal growths. Tumours are divided into two types: benign tumours and malignant tumours.

4.2 EXTRACTION PROCESSES OF PHYTOCHEMICALS

4.2.1 Solvent Extraction

Various phytoconstituents have been extracted using various solvents. To reduce the initial high-moisture content and allow for a longer storage life, the plant parts are quickly dried, ideally in the shade or in a controlled atmosphere at a low temperature (50–60°C). Mechanical grinders are used to grind the dried berries, and solvent extraction is used to remove the oil. The defatted substance is then extracted using a Soxhlet device, water, or alcohol (95% v/v), among other methods. The resulting alcoholic extract is then filtered, vacuum- or evaporatively concentrated, HCl (12*N*)-treated, and refluxed for at least 6 hours. Because saponins typically have high molecular weights, isolating them in their purest form may be difficult. The plant's components (tubers, roots, stems, leaves, and so on) are cleaned, cut into slices, and extracted for several hours in hot water or ethanol (95% v/v). The desired constituent is precipitated with ether after the resulting extract is filtered and vacuum-concentrated. Exhaustive extraction (EE) is commonly used to extract as many of the most active components with the highest biological activity as possible.

4.2.1.1 Few Examples of Anti-cancer Drugs Extracted by This Method

- *Catharanthus roseus* (Apocynaceae): Vindesine, vinorelbine, vinflunine
- *Podophyllum peltatum* and *Podophyllum emodi* (Berberidaceae): Etoposide, teniposide, Taxol, Taxotere
- *Camptotheca acuminate* (Nyssaceae): Topotecan, irinotecan, exatecan, LE–SN–38
- *Berberis amurensis* (Berberidaceae): Berbamine
- *Hydrastis canadensis* L., *Berberineeris* sp., and *Arcangelisia flava* (Ranunculaceae): Berberine
- *Tabebuia avellanedae* (Bignoniaceae): Beta-lapachone
- *Betula alba* (Bignoniaceae): Betulinic acid

4.2.2 Supercritical Fluid Extraction

This is the most technologically advanced extraction system. Gases, typically CO_2, are used in supercritical fluid extraction (SFE) by being compressed into a viscous liquid. The substance to be removed is then pushed through a cylinder containing this liquid. The liquid that is loaded with extract is then pushed into a chamber where the extract and gas are separated, with the gas being collected for further use. By adjusting the pressure and temperature at which one works, one can control and modify the solvent characteristics of CO_2. The flexibility SFE affords in identifying the elements you wish to extract from a specific material and the fact that your product has almost no solvent residues left in it are its two main benefits (CO_2 evaporates completely). The drawback of this technique is its high cost. There are numerous different gases and liquids that, when put under pressure, work incredibly well as extraction solvents.

4.2.2.1 Few Examples of Anti-cancer Drugs Extracted by This Method

- *Ochrosia borbonica*, *Excavatia coccinea*, and *Ochrosia elliptica* (Apocynaceae): Ellipticine
- *Rhizome of Rheum emodi (*Rhubarb) (Polygonaceae): Emodin
- *Amoora rohituka* and *Dysoxylum binectariferum* (Meliaceae): Flavopiridol
- *Cephalotaxus harringtonia*, *C. hainanensis*, and *C. sinensis* (Cephalotaxaceae): Harringtonine, homoharringtonine
- *Chinese herb, Dang gui*: Indirubin
- *Euphorbia peplus* L. (Euphorbiaceae): Ingenol-3-angelate

4.2.2.2 Coupled Supercritical Fluid Extraction–Supercritical Fluid Chromatography

In this method, a sample is first placed in the chromatographic system, where it is extracted with a supercritical fluid before being directly chromatographed.

4.2.2.2.1 Few Examples of Anti-cancer Drugs Extracted by This Method

- *Curcuma longa* (Zingiberaceae): Curcumin
- *Wikstroemia indica* (Thymelaeaceae): Daphnoretin
- *Lupinus* species, *Vicia faba*, *Glycine max*, and *Psoralea corylifolia* (Leguminosae): Daidzein, genistein

4.2.2.3 Coupled Supercritical Fluid Extraction–Gas Chromatography and Supercritical Fluid Extraction–LC

In this system, a sample is extracted with the use of a supercritical fluid, which is subsequently depressurized to deposit the extracted material in the intake portion or a column of a gas chromatographic system or a liquid chromatographic system, as appropriate. SFE has distinctive qualities such robust sample preparation, dependability, high yield, and reduced processing time. It also has the potential to be coupled with a variety of chromatographic techniques.

4.2.2.3.1 Few Examples of Anti-cancer Drugs Extracted by This Method

- *Salvia prionitis* Hance (Lamiaceae): Salvicine
- *Centaurea schischkinii* (Asteraceae): Schischkinnin
- *Centaurea Montana* (Asteraceae): Montamine
- *Aglaia foveolata* Panell (Meliaceae): Silvestrol
- *Tripterygium wilfordii* Hook F. (Celastraceae): PG490-88

4.2.3 Microwave-assisted Extraction

Microwave-assisted processing (MAP), a cutting-edge method of microwave-assisted solvent extraction, is used to extract high-value substances from natural sources, such as phytonutrients, components for nutraceuticals and functional foods, and pharmaceutical actives from biomass. The benefits of MAP technology over traditional solvent extraction methods include (1) improved products, increased purity of crude extracts, improved stability of marker compounds, and the potential to use less toxic solvents; and (2) decreased processing costs, increased recovery and purity of marker compounds, extremely high extraction rates, and decreased energy and solvent usage. In contrast to diffusion, microwave-derived extraction can achieve extremely rapid extraction rates and greater solvent flexibility. The microwave power and energy density, among other factors, can be adjusted to provide desired product qualities and enhance process economics. Excellent extracts can be produced from a wide range of substrates by customizing the process to optimize for commercial and financial considerations.

4.2.3.1 Few Examples of Anti-cancer Drugs Extracted by This Method

- *Euphorbia peplus* L. (Euphorbiaceae): Ingenol-3-angelate
- *Ipomoea batatas* (Convolvulaceae): 4-Ipomeanol
- *Iridaceaelatea pallasii* and *Iris kumaoensis* (Iridaceae): Irisquinone
- *Erythroxylum pervillei* (Erythroxylaceae): Pervilleines

4.2.4 Solid-phase Extraction

This involves the same principles that keep molecules on chromatographic stationary phases retaining solutes from a liquid medium onto a solid adsorbent. Similar to chromatographic media, these adsorbents are available as resins or beads that can be utilized in columns or in batches. They are frequently employed in the form of syringes, which are commercially available and contain medium that can be gently driven through with a plunger or by suction. The amount of medium in these syringes ranges from a few hundred milligrams to a few kilograms. Reverse-phase, normal-phase, and ion-exchange media are all examples of solid-phase extraction media. This sample purification technique uses adsorption onto a single-use solid-phase cartridge to concentrate and isolate the analyte from a solution of crude extracts. Typically, the analyte is kept on the stationary phase, cleaned, and then assessed with various mobile phases. Everything that is somewhat non-polar will bind if an aqueous extract is passed down a column filled with reverse-phase packing material, whereas everything polar will flow through.

4.3 ISOLATION OF PHYTOCHEMICALS

4.3.1 Liquid Chromatography

4.3.1.1 Preparative High-performance Liquid Chromatography

Preparative high-performance liquid chromatography (HPLC) can mainly be divided into two categories. One is standard preparative layer (PLC), which uses glass or plastic columns filled with poor-efficiency packing materials that have a huge particle size distribution and low pressure (usually under 5 bars). Preparative HPLC, a more contemporary variation of PLC, is becoming more and more common in the pharmaceutical sector.

Compounds are to be isolated or purified, but in analytical work, information about the sample is what is sought after. Preparative HPLC is more similar to analytical HPLC than standard PLC since it can perform more challenging separations more quickly, thanks to its better column efficiencies and faster solvent velocities. In analytical HPLC, resolution, sensitivity, and quick analysis time are crucial factors, but in preparative HPLC, both the level of solute purity and the quantity of chemical that can be produced in a given length of time, also known as throughput or recovery, are crucial. This is crucial in the modern pharmaceutical sector because novel products—natural and synthetic—must be released into the market as soon as possible. Spending less time on the synthesis conditions is achievable, thanks to the powerful purification process that is readily available.

4.3.1.2 Few Examples of Anti-cancer Drugs Extracted by This Method

- *Wikstroemia indica* (Thymelaeaceae): Daphnoretin
- *Psoralea corylifolia* (Leguminosae): Angelicin and *Vicia faba* (Fabaceae): Epicatechin
- *Amoora rohituka* and *Dysoxylum binectariferum* (Meliaceae): Flavopiridol
- *Cephalotaxus harringtonia*, *C. hainanensis*, and *C. sinensis* (Cephalotaxaceae): Harringtonine, homoharringtonine
- Chinese herb, *Dang gui*: Indirubin

4.3.1.3 Liquid Chromatography–Mass Spectroscopy

In several phases of drug development, liquid chromatography–mass spectroscopy (LC–MS) has emerged as the method of choice in the pharmaceutical sector. Recent innovations include liquid secondary ion mass spectroscopy, later laser mass spectroscopy with 600 MHz, which provides accurate determination of molecular weight proteins and peptides, as well as electrospray (ES), thermospray, and ion spray ionization techniques, which offer unique advantages of high detection sensitivity and specificity. This method can be used to find isotope patterns.

4.3.1.4 Liquid Chromatography–Nuclear Magnetic Resonance

One of the most effective and efficient approaches for the separation and structural elucidation of unknown compounds and mixtures, particularly for the structure elucidation of light- and oxygen-sensitive substances, is the coupling of chromatographic separation technology with nuclear magnetic resonance (NMR) spectroscopy. The processing in liquid chromatography–nuclear magnetic resonance (LC–NMR) improves speed and sensitivity of detection, and the online LC–NMR technique enables continuous registration of temporal changes as they arise in the chromatographic run automated data capture. The recent development of the three-dimensional approach and the pulsed field gradient technique in high-resolution NMR increases the applicability in structure elucidation and molecular weight data. The domains of pharmacokinetics, toxicology research, drug metabolism, and the drug discovery process can all benefit from these novel hyphenated approaches.

4.3.1.5 Few Examples of Anti-cancer Drugs Extracted by This Method

- *Centaurea schischkinii* (Asteraceae): Schischkinnin
- *Centaurea Montana* (Asteraceae): Montamine
- *Aglaia foveolata* Panell (Meliaceae): Silvestrol
- *Tripterygium wilfordii* Hook F. (Celastraceae): PG490-88

4.3.2 Gas Chromatography

4.3.2.1 Gas Chromatography–Fourier Transform Infrared Spectrometry

A powerful method for separating and identifying the constituents of various mixes is the combination of capillary column gas chromatographs and a Fourier transform infrared spectrometer.

4.3.2.2 Few Examples of Anti-cancer Drugs Extracted by This Method

- *Catharanthus roseus* (Apocynaceae): Vindesine, vinorelbine, vinflunine
- *Podophyllum peltatum* and *Podophyllum emodi* (Berberidaceae): Etoposide, teniposide, Taxol, Taxotere
- *Camptotheca acuminate* (Nyssaceae): Topotecan, irinotecan, Exatecan, LE–SN–38
- *Berberis amurensis* (Berberidaceae): Berbamine

4.3.2.3 Gas Chromatography–Mass Spectroscopy

Different types of rapid scan mass spectrometers can be directly interfaced with gas chromatography (GC) equipment. In most cases, the capillary column's flow rate is low enough for the column output to be fed directly into the MS' ionization chamber. Ion trap detectors (ITD) are the GC's most basic mass detector. By electron impact or chemical ionization, ions are produced from the eluted material in this instrument and stored in a radio frequency field. The trapped ions are subsequently released from the storage area and land on an electron multiplier detector. Controlled ejection makes it feasible to scan on the basis of mass-to-charge ratio. The ITD is significantly more affordable and smaller than quadrupole equipment. Numerous components that are present in natural and biological systems have been identified using GC–MS equipment.

4.3.2.4 Few Examples of Anti-cancer Drugs Extracted by This Method

- *Wikstroemia indica* (Thymelaeaceae): Daphnoretin
- *Lupinus* species, *Vicia faba*, *Glycine max*, and *Psoralea corylifolia* (Leguminosae): Daidzein, genistein

4.3.3 Supercritical Fluid Chromatography

A mix of gas and liquid chromatography that incorporates some of the greatest aspects of each is known as supercritical fluid chromatography (SFC). This method, a crucial third type of column chromatography, is starting to be used in a lot of commercial, government-regulated, and university labs. SFC is significant because it enables the separation and identification of a class of substances that neither gas nor liquid chromatography makes it easy to handle. These substances either lack a functional group that would enable detection by spectroscopic or electrochemical methods used in LC, or they are non-volatile or thermally labile, making GC processes inapplicable. SFC has been used on a wide range of substances, including organic goods, pharmaceuticals, foods, and insecticides.

4.3.3.1 Few Examples of Anti-cancer Drugs Extracted by This Method

- *Berberis amurensis* (Berberidaceae): Berbamine
- *Hydrastis canadensis* L., *Berberineeris* sp., and *Arcangelisia flava* (Ranunculaceae): Berberine
- *Tabebuia avellanedae* (Bignoniaceae): Betalapachone
- *Betula alba* (Bignoniaceae): Betulinic acid
- *Colchicum autumnale* and *Gloriosa superb* L. (Colchicaceae): Colchicine

4.3.4 Other Chromato-Spectrometric Studies

For establishing C–H bonds and determining connectivity between nearby protons, NMR techniques are used. Long-range heteronuclear correlations over multiple bonds are another use for INEPT. For the separation and determination of the structure of antifungal and antibacterial plant compounds, more and more techniques are being used, including thin layer chromatography (TLC), HPLC, HPLC coupled with ultraviolet (UV) photodiode array detection, LC–UV, LC–MS, ES, and LC–NMR. There are numerous chromatographic and spectroscopic methods for finding novel drugs from natural materials. The interpretation of spectra and the creation of chemical structures that match the spectral characteristics of bioactive substances produced from plants both have benefited from the introduction of computer modelling. The computer systems make use of spectrum features from 1H, 13C, 2D-NMR, IR, and MS. One can browse through libraries of spectra to compare them to full or incomplete chemical structures. High-throughput biological screening is combined with hyphenated chromatographic and spectroscopic techniques, which are potent analytical tools, to prevent the re-isolation of known compounds and to determine the structures of novel compounds. LC–UV–MS, LC–UV–NMR, LC–UV–ES–MS, and GC–MS are examples of hyphenated chromatographic and spectroscopic techniques.

4.3.4.1 Few Examples of Anti-cancer Drugs Extracted by This Method

- *Wikstroemia indica* (Thymelaeaceae): Daphnoretin
- *Lupinus species*, *Vicia faba*, *Glycine max*, and *Psoralea corylifolia* (Leguminosae): Daidzein, genistein
- *Ochrosia borbonica*, *Excavatia coccinea*, *Ochrosia elliptica* (Apocynaceae): Ellipticine
- *Rhizome of Rheum emodi* (Rhubarb) (Polygonaceae): Emodin

4.4 PHYTOCHEMICALS IN CANCER TREATMENT

Phytochemicals and derivatives present in plants are promising options to improve treatment efficiency in cancer patients and decrease adverse reactions. Several of these phytochemicals are naturally occurring biologically active compounds with anti-tumour activity. The development of effective and side effect–free phytochemical-based anti-cancer therapy begins with the testing of natural extracts (from dry/wet plant material) for potential anti-cancer biological activity, which is followed by the purification of active phytochemicals using bioassay-guided fractionation and testing for *in vitro* and *in vivo* effects.

The chapter attempted to gather information specifically about anti-cancer phytochemicals that have been evaluated at preclinical and clinical levels, as well as those that are currently available on the market. We included phytochemicals with reported *in vivo* activity in the preclinical section. This chapter also discusses phytochemicals that have been evaluated at the preclinical level, as well as some phytochemicals that are currently being studied in clinical trials, as well as a brief overview of the currently available plant-based anti-cancer drugs.

4.4.1 Phytochemicals with Anti-cancer Properties

Phytochemicals have significant anti-tumour potential, according to scientific evidence. From 1940 to 2014, approximately half of all approved anti-cancer drugs were derived directly or indirectly from natural products. In this chapter, we will look at some of the most effective anti-cancer phytochemicals. These phytochemicals have been evaluated for anti-cancer efficacy *in vitro* and *in vivo*.

They have complementary and overlapping mechanisms that slow down the carcinogenic process by scavenging free radicals, suppressing malignant cell survival and proliferation, and decreasing tumour invasiveness and angiogenesis. They act on a diverse set of molecular targets and signalling pathways, including membrane receptors, kinases, downstream tumour-activator or tumour-suppressor proteins, transcriptional factors, microRNAs (miRNAs), cyclins, and caspases.

4.4.2 Phytochemicals Isolated from Plants Possessing Anti-cancer Activity

A careful use of preclinical screening models can produce potential lead compounds for the development of anti-cancer drugs with extensive data on preliminary efficacy, toxicity, pharmacokinetic, and safety information that can help determine whether a molecule should be taken further for clinical trials. In the framework of this chapter, a wealth of information regarding the preclinical effectiveness of several phytochemicals has been gathered.

4.4.2.1 6-Shogaol

A small bioactive ingredient identified from ginger is called 6-shogaol (*Zingiber officinale* Roscoe). The growth of NCI-H1650 lung cancer cells was significantly inhibited by 6-shogaol (10 mg/kg) in a nude mice model of non-small cell lung cancer (NSCLC). This was accompanied by decreased cell proliferation and increased apoptosis, as shown by a reduction in Ki-67-positive cells and an increase in terminal deoxynucleotidyl transferase deoxyuridine triphosphate nick-end labelling (through direct targeting of Akt1 and Akt2, 6-shogaol inhibited Akt signalling at the *in vitro* level. Intraperitoneal treatment of 6-shogaol (100 mg/kg body weight) decreased tumour weight in a syngeneic FVB/N mouse model of prostate cancer, which was correlated with a decline in pSTAT3Y705, cyclin D1, and surviving levels.

4.4.2.2 Allicin

The effects of allicin, one of the primary organic allyl sulphur components of garlic (*Allium sativum*, Amaryllidaceae), on cholangiocarcinoma were investigated (CCA). Allicin (10 mg/kg) effectively slowed the growth of human liver bile duct carcinoma in BALB/c nude mice models of CCA (HuCCT-1). The *in vitro* molecular analysis revealed that allicin (20 M) decreased HuCCT-1 cell motility, invasion, and epithelial–mesenchymal transition (EMT) via lowering the levels of matrix metalloproteinase (MMP)-2 and (MMP)-9 and decreasing the activity of the STAT3 signalling pathway. Allicin also inhibited proliferation by upregulating Bcl-2-associated X (Bax) protein and activating the caspase cascade, inducing apoptosis, and downregulating the expression of STAT3-related proteins such B-cell lymphoma 2 (Bcl-2). Later research revealed that allicin (5 M) changed the TIMP/MMP balance by lowering the activity of the PI3K/AKT signalling pathway, greatly suppressing lung adenocarcinoma A549 and H1299 cell adhesion, invasion, and migration.

4.4.2.3 Alpinumisoflavone

The pyranoisoflavone alpinumisoflavone (AIF) is found in the plant *Derris eriocarpa* (Leguminosae). AIF (40 mg/kg) inhibited the growth and metastasis of 786-O human clear cell renal cell cancer (ccRCC) cells when they were transplanted into BALB/c nude mice. By inhibiting Akt signalling (i.e. lowering RLIP76 expression and the p-Akt/t-Akt ratio), miR-101 expression was increased, which led to the inhibitory impact. The expression of the antioxidant molecules NQO-1 and HO-1, which are driven by the nuclear transcription factor Nrf2, as well as reactive oxygen species (ROS) formation, DNA damage apoptosis, and cell cycle arrest, are all reportedly increased by AIF in esophageal squamous cell carcinoma (ESCC).

4.4.2.4 Andrographolide

A bicyclic diterpenoid lactone called andrographolide was discovered in the *Andrographis paniculata* plant (Acanthaceae). It was discovered that andrographolide inhibits tumour growth by preventing tumour adaption to hypoxic conditions. The hypoxia-inducible factor (HIF)-1a activity and its upstream PI3k/AKT/mTOR pathway were inhibited, which is why andrographolide (100 mg/kg) had the desired effect.

4.4.2.5 Apigenin

Natural flavonoid apigenin (APG), which is found in fruits and vegetables, has a variety of anti-cancer effects. APG (5 mg/kg) inhibited tumour growth in an athymic nude mouse xenograft with human chondrosarcoma Sw1353 cells, which was accompanied by a decrease in Ki67 expression and induction of apoptosis. APG controlled the expression of Bcl-2 family proteins at the molecular level and triggered the caspase cascade to cause G2/M phase arrest and apoptosis. APG (3 mg/kg) was used in a different investigation to inhibit the dipeptidyl peptidase IV (DPPIV) enzyme and lessen the growth and metastasis of NSCLC xenografts. APG modulated the EMT and the capacity for invasion of both the EGFR positive and EGFR negative NSCLC cells by suppressing the snail/slug signalling and downregulating the DPPIV enzyme, according to *in vitro* mechanistic studies. APG's effectiveness was shown to be increased in several preclinical tests when it was coupled with other chemotherapeutic drugs or placed onto nanocarriers.

4.4.2.6 Baicalein and Baicalin

The naturally occurring flavonoids and active ingredients of *Scutellaria baicalensis* are called baicalein and baicalin (Lamiaceae). Baicalein (50 mg/kg) and baicalin (50 mg/kg) reduced tumour growth and promoted apoptosis in NOD-scid IL2Rg null (NSG) mice xenograft with human colon cancer HCT116 cells. *In vivo*, it reduced the expression of the human telomerase reverse transcriptase (hTERT) and inhibited the signalling pathways for extracellular receptor kinase (ERK), mitogen-activated protein kinase (MAPK), and p38. The intraperitoneal treatment of baicalin (50 mg/kg) in a different investigation in a nude mice model of colon cancer decreased tumour growth by inhibiting the production of c-Myc and oncomiRs microRNAs to promote apoptosis. Additionally, baicalein (50 mg/kg) decreased the growth of the tumour by enhancing apoptosis and reducing tumour angiogenesis when combined with docetaxel (10 mg/kg).

4.4.2.7 Curcumin

From the plant *Curcuma longa*, curcumin (phytopolylphenol) is a phytochemical (Zingiberaceae). Curcumin has been shown to have anti-cancer properties through modification of various signalling and gene expression regulatory mechanisms, according to several research. When mice were subcutaneously injected with human A375 melanoma cells, curcumin slowed the growth of the tumours. Through processes such as cell cycle arrest, autophagy, and downregulation of the PI3K/AKT/mTOR/P70S6K pathway—a crucial intracellular signalling system linked to cell survival and death—studies showed that curcumin suppressed the proliferation of melanoma cells.

4.4.2.8 Decursin and Decursinol

Coumarins, specifically decursin and decursinol, are extracted from the dried roots of *Angelica gigas* Nakai. In rats and people, decursin is quickly and completely transformed to decursinol. Decursinol (4.5 mg/mouse) inhibits tumour growth and lung metastasis in SCID-NSG mice xenografted with human prostate cancer LNCaP/AR-Luc cells overexpressing the wild-type androgen receptors (AR).

4.4.2.9 Dicumarol

The natural anticoagulant dicumarol (DIC) is produced by bacterial action in rotten sweet clover hay and is derived from coumarin (*Melilotus officinalis*, Fabaceae). DIC (30 mg/kg) effectively slowed the development of SKOV3 ovarian cancer cells in a BALB/c nude mice xenograft model. The *in vitro* molecular mechanistic studies suggested that DIC increased the level of ROS, switched the metabolism of glucose from aerobic glycolysis to oxidative phosphorylation, inhibited the kinase activity of pyruvate dehydrogenase kinase 1 (PDK1), attenuated the mitochondrial membrane potential (MMP), induced apoptosis, and decreased cell viability of SKOV3 cells. When female fertility preservation is a priority, it is significant that DIC (32 mg/kg) was proven to be safe for ovarian tissues and developing oocytes; this suggests the importance of DIC as a potential anti-cancer drug.

4.4.2.10 Epigallocatechin

By triggering apoptosis and obstructing proliferation of human breast cancer MDA-MB-231 cells in a nude mice model, epigallocatechin (EGCG), a key catechin present in green tea, successfully delayed the formation of tumours and decreased tumour burden. In a different study, EGCG reduced the rise in the oxidative stress-derived DNA damage marker 8-hydroxydeoxyguanosine (8-OH-dGuo) levels in rat lung DNA to prevent the development of lung tumours caused by the nitrosamine (NNK) compound.

4.4.2.11 Emodin

The rhizome and root of *Rheum palmatum* L. are the sources of the anthraquinone derivative emodin (Polygonaceae). Emodin (50 mg/kg) reduced the development of human lung epithelial (A549) cells in BALB/c nude mice via causing ER stress-dependent apoptosis. Emodin stimulated ER stress and TRIB3/nuclear factor-κB signalling, according to the *in vitro* molecular mechanism. Emodin inhibited macrophage infiltration and M2-like polarization in mice with EO771 or 4T1 breast cancers, which was followed by enhanced T-cell activation and decreased tumour angiogenesis. Emodin dramatically elevated inhibitory histone H3 lysine 27 tri-methylation (H3K27m3) on the promoters of M2-related genes in tumour-associated macrophages and reduced IRF4, STAT6, and C/EBPb signalling at the molecular level. Emodin inhibited tumour growth and promoted apoptosis in BALB/c nude mice that were xenografted with human hepatocellular carcinoma SMMC-7721 cells. This was accompanied by an increase in ERK and p38 phosphorylation and a decrease in p-JNK expression.

4.4.2.12 Genistein

Soy beans contain genistein, an isoflavone that naturally occurs and has characteristics similar to those of oestrogen. In the azoxymethane (AOM)-induced rat colon cancer model, genistein (140 mg/kg) therapy reduced the overall number of aberrant crypts by inhibiting aberrant nuclear accumulation of β-catenin and suppressing WNT signalling genes. Genistein (0.4 mg/kg) was intraperitoneally given for 28 days in an athymic BALB/c nu/nu mouse xenograft with the human leukaemia cell line HL-60. This dramatically decreased the tumour weight without changing the body weight. Through ROS-mediated ER stress, genistein caused G2/M phase arrest and death of HL-60 cells *in vitro*. This increased $Ca^{2}+$ generation and lowered mitochondrial membrane potential. Increased expression of proteins related with ER stress (IRE1a, calpain 1, GRP78, GADD153, caspase-7, caspase-4, and ATF6a) and proteins associated with apoptosis (Bax, PARP cleavage, caspase-9, caspase-3, Bcl2, and Bid) was the cause of the observed effect at the molecular level.

4.4.2.13 Gingerol

The main phenolic ingredient in ginger's rhizomes is called gingerol (*Z. officinale* Roscoe). Gingerol (5 mg/kg) treatment caused caspase-3 activation and inhibited the growth of orthotopic tumours in a syngeneic mouse model of spontaneous breast cancer metastasis as well as the metastasis of mouse brain metastatic 4T1Br4 mammary tumour cells to multiple organs, including the lung, bone, and brain. Joo and his co-workers have reported that gingerol inhibits lung metastases, MDA-MB-231 human breast cancer cell proliferation, and invasion by p38MAPK, Akt, and the epidermal growth factor receptor.

4.4.2.14 Glycyrrhizin

The main bioactive substance present in the licorice roots of the tiny leguminous plant *Glycyrrhiza glabra* L. is glycyrrhizin (GA). GA (135 mg/kg) lowered thromboxane synthase (TxAS) and proliferating cell nuclear antigen (PCNA) expression in athymic BALB/c nude mice xenografted with human lung adenocarcinoma A549 cells stably transfected with TxA2 receptor (TPa) by decreasing TxA2 pathway. More recent research revealed that GA (100 mg/kg) reduced the level of high mobility group box 1 (HMGB1) and blocked the JAK/STAT signalling pathway, which in turn reduced the growth of NSCLC in patient-derived xenograft (PDX) mice.

4.4.2.15 Hispidulin

A phenolic flavonoid called hispidulin is present in a variety of plant materials, including *Saussurea involucrata* Kar. (Asteraceae). By upregulating cleaved caspase-3 expression and downregulating Sphk1 activity, hispidulin (20 mg/kg) intraperitoneally reduced the growth of the Caki-2 (human clear cell renal cell carcinoma) tumour and lung metastasis in athymic BALB/c nu/nu mice model. Similar to this, another study found that hispidulin (20 mg/kg) successfully reduced lung metastasis and tumour growth in human hepatocellular carcinoma Bel7402 cell xenografts by upregulating PPARg expression and AMPK, JNK, and ERK protein phosphorylation levels. A synthetic resveratrol counterpart with enhanced photosensitivity and stability is called HS-1793. Using a nude mouse (homozygous for the nu mutation) breast cancer model.

4.4.2.16 Licochalcone A

A phenol chalconoid called licochalcone A (LicA) was discovered in the roots of *Glycyrrhiza* species. LicA (20 mg/kg) reduced the development of the human cervical cancer cell SiHa tumour in the athymic BALB/c nu/nu mice model by blocking the PI3K/Akt/mTOR signalling pathway and inducing apoptosis. By lowering the levels of cyclins and cyclin-dependent kinases, LicA caused cell cycle arrest in the G0/G1 and G2/M phases in athymic nude mice orthotopic or subcutaneous xenograft with human glioma U87 cells. Most recently, LicA was demonstrated to inhibit Akt signalling pathway–mediated tumour glycolysis in gastric cancer via inhibiting hexokinase 2–mediated tumour glycolysis.

4.4.2.17 Nimbolide

A triterpene called nimbolide is generated from the neem tree's leaves and blossoms (*Azadirachta indica*). Nimboide (5 mg/kg) reduced the pancreatic cancer HPAC cell growth and metastasis in an athymic nu/nu mice model via triggering apoptosis. Nimboide increased ROS production, inhibited proliferation (through decreased PI3K/AKT/mTOR and ERK signalling), and metastasized (through decreased EMT, invasion, migration, and colony-forming abilities) via mitochondrial-mediated apoptotic cell death, according to *in vitro* molecular mechanism studies. Nimbolide may play an epigenetic role in controlling autophagy and apoptosis in human breast cancer cells, according to a recent *in vitro* study.

4.4.2.18 Physapubescin B

A steroidal compound called physapubescin B was purified from *Physalis pubescens* L. (Solanaceae). Physapubescin B (50 mg/kg) inhibited the growth of the PC3 tumour in nude mouse models with prostate cancer xenografts by decreasing the expression levels of Ki-67, Cdc25C, and full length PARP and increasing the number of apoptotic cells in the tumour tissue (Ding et al., 2015). Additionally, physapubescin (30 mg/kg) reduced vimentin protein expression and prevented *in vivo* angiogenesis in renal cell carcinoma 786-O cells.

4.4.2.19 Pterostilbene

A naturally occurring resveratrol derivative derived from grapes is called pterostilbene (*Vitis vinifera*, Vitaceae). Pterostibene (100 or 200 mg/kg) dramatically reduced EC109 tumour growth, cell adhesion, migration, and intracellular glutathione (GSH) levels while raising the apoptotic index, caspase-3 activity, and ROS levels in an athymic nude mouse oesophageal cancer model. Pterostibene (30 mg/kg) also significantly slowed tumour growth in an athymic nude mouse model of diffuse large B-cell lymphoma, decreased MMP, increased cellular apoptotic index, and increased ROS levels, which caused S-phase arrest in the cell cycle. Also, it was shown that pterostilbene (30 mg/kg) and megestrol acetate (10 mg/kg) effectively inhibited the growth of the HEC-1A tumour in a mouse model of endometrial cancer when compared to pterostilbene or megestrol acetate alone. The aforementioned combination reduced oestrogen receptor expression and the activity of the ERK and STAT3 signalling pathways *in vitro*.

4.4.2.20 Resveratrol

A polyphenolic phytoalexin is resveratrol (stilbenoid). Numerous studies have demonstrated that resveratrol inhibits the growth of a wide range of tumour cells, including those in the liver, colon, prostate, breast, and lung. When animals with highly metastasizing Lewis lung carcinoma tumours were given resveratrol, the tumour development and metastasis to the lung were greatly reduced. According to the findings, resveratrol's anti-cancer and anti-metastatic properties may be due to its ability to suppress DNA synthesis, neovascularization, and angiogenesis. Resveratrol decreased the frequency and number of tumours in a breast cancer model created by 7,12-dimethylbenz[*a*]anthracene (DMBA), while also lengthening the latency period. In the same study, resveratrol was able to inhibit nuclear factor-κB activation, which controls the gene production of matrix metalloproteinase-9 and cyclooxygenase-2.

4.4.2.21 Sulforaphane

One of the isothiocyanate group of organosulphur compounds is sulforaphane (SFN). In addition to boosting the anti-cancer effectiveness of other anti-proliferative drugs like paclitexal, SFN exerts its anti-cancer effects by altering important signalling pathways such as the induction of apoptosis, suppression of cell cycle progression, and inhibition of angiogenesis. Compared to SFN or paclitexal alone, the addition of SFN and paclitexal to Barrett esophageal adenocarcinoma (BEAC) cells greatly boosted apoptotic cell death. SFN also detected a significant decrease in tumour volume in mice with severe combined immunodeficiency (SCID) that had received subcutaneous injections of BEAC cells.

4.4.2.22 Thymol

Thyme (*Thymus vulgaris*) and oregano contain thymol, a transient receptor potential ankyrin subtype 1 (TRPA1) channel agonist (*Origanum vulgare*). Thymol (4.3 mmol/L) intratumour injection in oral squamous cell carcinoma Cal27 and HeLa-derived mouse xenografts decreased tumour volume by reducing cell proliferation and triggering apoptosis, as shown by Ki-67 staining and TUNEL tests, respectively. According to *in vitro* molecular mechanism investigations, thymol caused the potential of the mitochondrial membrane to depolarize in order to cause apoptosis.

4.4.2.23 Thymoquinone

Black cumin (*Nigella sativa*, Ranunculaceae) seed oil contains the active ingredient thymoquinone (2-isopropyl-5-methyl-1,4-benzo-quinone, or TQ). TQ (10 mg/kg) reduced tumour weight and size in BALB/c athymic nude mice via triggering apoptosis and preventing STAT3 activation in human gastric cancer cells. JAK2 and c-Src activity decreased along with the downregulation of STAT3 activation. The potential of TQ in adjuvant therapy with other chemotherapeutic drugs has recently been revealed by preclinical research. In a different experiment, TQ in combination with melatonin significantly reduced the tumour size, induced tumour cell death, reduced VEGF expression, and activated the anti-cancer immune response by raising serum interferon INF-γ levels in BALB/c mice transplanted with mouse epithelia breast cancer EMT6/P cell line.

4.4.2.24 Ursolic Acid

A natural terpene molecule called ursolic acid (UA) can be found in many different types of plants. Recent research have suggested using UA as a cancer chemosensitizer to common chemotherapeutic medicines due to its well-known anti-cancer properties. In one study, it was discovered that UA, which inhibits the tumour and raises survival rates, improves the therapeutic effects of oxaliplatin in a mouse model of CRC. According to an *in vitro* mechanistic investigation, treatment of CRC cells with UA plus oxaliplatin greatly reduced the ability of the cells to proliferate, increased the generation of apoptosis and ROS, and significantly reduced the expression of drug-resistant genes. By targeting caspases and p53 and downregulating Bcl-2 and cIAP, the UA nanoparticles reduced tumour growth and caused the death of cervical cancer cells.

4.4.2.25 Withaferin A

Withania somnifera contains the steroidal lactone withaferin A (WA) (Solanaceae). Oral treatment of WA (5 mg/kg) prevented the development of microvessels and tumour growth in human colorectal carcinoma (HCT-116) cells overexpressing AKT in a nude mouse model of colorectal cells (CRC). WA decreased EMT indicators, which at the *in vitro* level in AKT overexpressing HCT-116 cells prevented cell proliferation, migration, and invasion (snail, slug, β-catenin, and vimentin). Another study found that withaferin-A (2 mg/kg) intraperitoneal injection prevented the growth of CRC via preventing STAT3 activation caused by interleukin-6. Similar to this, in another study, oral administration of WA (4 mg/kg) prevented diethylnitrosamine (DEN)-induced hepatocellular carcinoma (HCC) and HepG2-xenografts in C57BL/6 mice by increasing ERK, RSK, ELK1, and DR5 levels and lowering Ki67 expression. According to *in vitro* molecular mechanism studies, WA raised the phosphorylation of ERK and p38, which in turn elevated the phosphorylation of p90-ribosomal S6 kinase (RSK) and concurrently activated ETS-like transcription factor-1 (ELK1) and death receptor protein-5 (DR5)

4.5 CONCLUSIONS AND FUTURE PROSPECTS

Although there hasn't been much research on specific phytochemicals in foods and how they affect illness risks, there is enough evidence from the link between foods high in phytochemicals and disease risks to strongly suggest that eating foods high in these components may help prevent diseases. It is unknown, nevertheless, whether the health advantages are due to specific phytochemicals, how different phytochemicals interact with one another, how much fibre is present in plant foods, or how phytochemicals interact with the vitamins and minerals present in the same foods. Consuming fruits, vegetables, and whole grains, as well as eating habits that place an emphasis on these items, like the Mediterranean diet, has been linked to a lower risk of developing numerous cancers, including breast, lung, and colon cancer. A daily increase of three servings of whole grains is linked to a 17% decreased risk of colorectal cancer. Studies that examined the consumption of phytochemicals discovered a connection between lower cancer risks. Only particular flavonoid groupings, according to one study, were linked to a lower risk of breast cancer. These studies indicated that compared to foods high in phytochemicals, the lowered risk of cancer wasn't as significant for specific phytochemicals. A lower incidence of prostate, lung, breast, and colon cancer has been linked to eating cruciferous vegetables, including broccoli, cabbage, and cauliflower. It is thought that the isothiocyanate phytochemicals in cruciferous vegetables, particularly the well-researched sulforaphane in broccoli, provide some level of protection. The discussion makes it very evident that phytochemicals are crucial in the fight against numerous illnesses such as cancer, diabetes, cardiovascular disease, and other illnesses. Today, the world is facing problems in finding a successful cancer treatment. Nearly all nations are conducting numerous studies on cancer prevention and treatment to save the lives of billions of people.

LIST OF ABBREVIATIONS

CDK6	Cyclin-dependent kinase 6
CE–DAD	Capillary electrophoresis–diode array detection
CNS	Central nervous system
DPPH	2,2-Diphenyl-1-picryl-hydrazyl-hydrate
EE	Exhaustive extraction
ES	Electrospray
GC–MS	Gas chromatography–mass spectroscopy
GLC	Gas–liquid chromatography
HM	Herbal medicines
HPLC	High-performance liquid chromatography
HPLC–DAD	High-performance liquid chromatography–diode array detection
HPLC–MS	High-performance liquid chromatography–mass spectroscopy
HPTLC	High-performance thin liquid chromatography
HPV	Human papillomavirus
IR	Infrared
ITD	Ion trap detectors
LC–MS	Liquid chromatography–mass spectroscopy
LC–NMR	Liquid chromatography–nuclear magnetic resonance
LC–UV	Liquid chromatography–ultraviolet
LC–UV–ES–MS	Liquid chromatography–ultraviolet–electrospray–mass spectroscopy
LC–UV–MS	Liquid chromatography–ultraviolet–mass spectroscopy
LC–UV–NMR	Liquid chromatography–ultraviolet–nuclear magnetic resonance
MAP	Microwave-assisted processing
MAPK	Mitogen-activated protein kinase
MS	Mass spectroscopy
NADPH	Nicotinamide adenine dinucleotide phosphate
NF-κB	Nuclear factor kappa B
NMR	Nuclear magnetic resonance
PAL	Phenylalanine
PLC	Preparative layer of chromatography
SFC	Supercritical fluid chromatography

SFE	Supercritical fluid extraction
SFE–GC	Supercritical fluid extraction–gas chromatography
SFE–LC	Supercritical fluid extraction–liquid chromatography
TCM	Traditional Chinese medicine
TLC	Thin layer chromatography

BIBLIOGRAPHY

Abdullaev, F. I., & Espinosa-Aguirre, J. J. (2004). Biomedical properties of saffron and its potential use in cancer therapy and chemoprevention trials. *Cancer Detection and Prevention*, *28*, 426–432.

Acharya, A., Das, I., Singh, S., & Saha, T. (2010). Chemopreventive properties of indole-3-carbinol, diindolylmethane and other constituents of cardamom against carcinogenesis. *Recent Patents on Food, Nutrition & Agriculture*, *2*, 166–177.

Amin, A., Hamza, A. A., Bajbouj, K., Ashraf, S. S., & Daoud, S. (2011). Saffron: A potential target for a novel anticancer drug against hepatocellular carcinoma. *Hepatology*, *54*, 857–867.

Angelini, A., Di Ilio, C., Castellani, M. L., Conti, P., & Cuccurullo, F. (2010). Modulation of multidrug resistance *p*-glycoprotein activity by flavonoids and honokiol in human doxorubicin-resistant sarcoma cells (MES-SA/DX-5): Implications for natural sedatives as chemosensitizing agents in cancer therapy. *Journal of Biological Regulators and Homeostatic Agents*, *24*, 197–205.

Arai, Y., Watanabe, S., Kimira, M., Shimoi, K., Mochizuki, R., & Kinae, N. (2000). Dietary intakes of flavonols, flavones and isoflavones by Japanese women and the inverse correlation between quercetin intake and plasma LDL cholesterol concentration. *Journal of Nutrition*, *130*, 2243–2250.

Aung, H. H., Wang, C. Z., Ni, M., Fishbein, A., Mehendale, S. R., Xie, J. T., Shoyama, C. Y., & Yuan, C. S. (2007). Crocin from *Crocus sativus* possesses significant anti-proliferation effects on human colorectal cancer cells. *Experimental Oncology*, *29*, 175–180.

Bakshi, H., Sam, S., Rozati, R., Sultan, P., Islam, T., Rathore, B., Lone, Z., Sharma, M., Triphati, J., & Saxena, R. C. (2010). DNA fragmentation and cell cycle arrest: A hallmark of apoptosis induced by crocin from Kashmiri saffron in a human pancreatic cancer cell line. *Asian Pacific Journal of Cancer Prevention*, *11*, 675–679.

Bathaie, S. Z., & Mousavi, S. Z. (2010). New applications and mechanisms of action of saffron and its important ingredients. *Critical Reviews in Food Science and Nutrition*, *50*, 761–786.

Bradlow, H. L., & Zeligs, M. A. (2010). Diindolylmethane (DIM) spontaneously forms from indole-3-carbinol (13C) during cell culture experiments. *In Vivo*, *24*, 387–391.

Calderon-Montano, J. M., Burgos-Moron, E., Perez-Guerrero, C., & Lopez-Lazaro, M. (2011). A review on the dietary flavonoid kaempferol. *Mini-Reviews in Medicinal Chemistry*, *11*, 298–344.

Canene-Adams, K., Lindshield, B. L., Wang, S., Jeffery, E. H., Clinton, S. K., & Erdman, J. W., Jr. (2007). Combinations of tomato and broccoli enhance antitumor activity in dunning r3327-h prostate adenocarcinomas. *Cancer Research*, *67*, 836–843.

Chaimbault, P. (2014). The modern art of identification of natural substances. In C. Jacob, G. Kirsch, A. J. Slusarenko, P. G. Winyard, & T. Burkholz (Eds.), *Recent advances in redox active plant and microbial products* (pp. 31–94). Springer.

Christophoridou, S., Dais, P., Tseng, L. H., & Spraul, M. (2005). Separation and identification of phenolic compounds in olive oil by coupling high performance liquid chromatography with postcolumn solid-phase extraction to nuclear magnetic resonance spectroscopy (LC-SPE-NMR). *Journal of Agricultural and Food Chemistry*, *53*, 4667–4679.

Chryssanthi, D. G., Dedes, P. G., Karamanos, N. K., Cordopatis, P., & Lamari, F. N. (2011). Crocetin inhibits invasiveness of MDA-MB-231 breast cancer cells via down regulation of matrix metalloproteinases. *Planta Medica*, *77*, 146–151.

Conforti, F., Sosa, S., Marrelli, M., Menichini, F., Statti, G. A., Uzunov, D., Tubaro, A., Menichini, F., & Loggia, R. D. (2008). *In vivo* anti-inflammatory and *in vitro* antioxidant activities of Mediterranean dietary plants. *Journal of Ethnopharmacology*, *116*, 144–151.

Cui, Y., Morgenstern, H., Greenland, S., Tashkin, D. P., Mao, J. T., Cai, L., Cozen, W., Mack, T. M., Lu, Q. Y., & Zhang, Z. F. (2008). Dietary flavonoid intake and lung cancer: A population-based case-control study. *Cancer*, *112*, 2241–2248.

Daffre, S., Bulet, P., Spisni, A., Ehret-sabatier, L., Rodrigues, E. G., & Travassos, L. R. (2008). Bioactive natural peptides. *Studies in Natural Products Chemistry*, *35*, 597–691.

Dai, J., & Mumper, R. J. (2010). Plant phenolics: Extraction, analysis and their antioxidant and anticancer properties. *Molecules*, *15*, 7313–7352.

Das, I., Das, S., & Saha, T. (2010). Saffron suppresses oxidative stress in DMBA-induced skin carcinoma: A histopathological study. *Acta Histochemica*, *112*, 317–327.

Emam, S. S., & Abd El-Moaty, H. I. (2009). Glucosinolates, phenolic acids and anthraquinones of *Isatis macrocarpa* Boiss and *Pseuderucaria clavate* (Boiss & Reut.) family: Cruciferae. *Journal of Applied Sciences Research*, *5*, 2315–2322.

Gacche, R. N., Shegokar, H. D., Gond, D. S., Yang, Z., & Jadhav, A. D. (2011). Evaluation of selected flavonoids as antiangiogenic, anticancer, and radical scavenging agents: An experimental and *in silico* analysis. *Cell Biochemistry and Biophysics*, *61*, 651–663.

Geraets, L., Haegens, A., Brauers, K., Haydock, J. A., Vernooy, J. H., Wouters, E. F., Bast, A., & Hageman, G. J. (2009). Inhibition of LPS-induced pulmonary inflammation by specific flavonoids. *Biochemical and Biophysical Research Communications*, *382*, 598–603.

Giovannucci, E., Ascherio, A., Rimm, E. B., Stampfer, M. J., Colditz, G. A., & Willett, W. C. (1995). Intake of carotenoids and retinol in relation to risk of prostate cancer. *Journal of the National Cancer Institute*, *87*, 1767–1776.

Greer, J. B., Modugno, F., Allen, G. O., & Ness, R. B. (2005). Androgenic progestins in oral contraceptives and the risk of epithelial ovarian cancer. *Obstetrics and Gynecology*, *105*, 731–740.

Harborne, J. B., & Williams, C. A. (1992). Advances in flavonoid research since 1992. *Phytochemistry*, *55*, 481–504.

Howitz, K. T., Bitterman, K. J., Cohen, H. Y., Lamming, D. W., Lavu, S., Wood, J. G., Zipkin, R. E., Chung, P., Kisielewski, A., Zhang, L. L., Scherer, B., & Sinclair, D. A. (2003). Small molecule activators of sirtuins extend *Saccharomyces cerevisiae* lifespan. *Nature*, *425*, 191–196.

Jeong, C. H., Bode, A. M., Pugliese, A., Cho, Y. Y., Kim, H. G., Shim, J. H., Jeon, Y. J., Li, H., Jiang, H., & Dong, Z. (2009). [6]-Gingerol suppresses colon cancer growth by targeting leukotriene A4 hydrolase. *Cancer Research*, *69*, 5584–5591.

Jiang, Z., Kempinski, C., & Chappell, J. (2016). Extraction and analysis of terpenes/terpenoids. *Current Protocols in Plant Biology*, *1*, 345–358.

Kahkonen, M. P., Hopia, A. I., Vuorela, H. J., Rauha, J. P., Pihlaja, K., Kujala, T. S., & Heinonen, M. (1999). Antioxidant activity of plant extracts containing phenolic compounds. *Journal of Agricultural and Food Chemistry*, *47*, 3954–3962.

Kaskiw, M. J., Tassotto, M. L., Mok, M., Tokar, S. L., Pycko, R., Thng, J., & Jiang, Z. H. (2009). Structural analogues of diosgenyl saponins: Synthesis and anticancer activity. *Bioorganic & Medicinal Chemistry*, *17*, 7670–7679.

Khan, N., Afaq, F., Khusro, F. H., Adhami, V. M., Suh, Y., & Mukhtar, H. (2012). Dual inhibition of phosphatidylinositol 3-kinase/Akt and mammalian target of rapamycin signaling in human non-small cell lung cancer cells by a dietary flavonoid fisetin. *International Journal of Cancer*, *130*, 1695–1705.

Kim, J. M., Kim, J. S., Yoo, H., Choung, M. G., & Sung, M. K. (2008). Effects of black soybean [*Glycine max* (L.) Merr.] seed coats and its anthocyanidins on colonic inflammation and cell proliferation *in vitro* and *in vivo*. *Journal of Agricultural and Food Chemistry*, *56*, 8427–8433.

Kim, J. E., Kwon, J. Y., Seo, S. K., Son, J. E., Jung, S. K., Min, S. Y., Hwang, M. K., Heo, Y. S., Lee, K. W., & Lee, H. J. (2010). Cyanidin suppresses ultraviolet B-induced COX-2 expression in epidermal cells by targeting MKK4, MEK1, and raf-1. *Biochemical Pharmacology*, *79*, 1473–1482.

Kim, Y. S., & Milner, J. A. (2005). Targets for indole-3-carbinol in cancer prevention. *Journal of Nutritional Biochemistry*, *16*, 65–73.

Kinghorn, A., Farnsworth, N., Soejarto, D., Cordell, G., Swanson, S., Pezzuto, J., Wani, M., Wall, M., Oberlies, N., Kroll, D., Kramer, R., Rose, W., Vite, G., Fairchild, C., Peterson, R., & Wild, R. (2003). Novel strategies for the discovery of plant derived anticancer agents. *Pharmaceutical Biology*, *41*, 53–67.

Lee, H. S., Seo, E. Y., Kang, N. E., & Kim, W. K. (2008). [6]-Gingerol inhibits metastasis of MDA MB-231 human breast cancer cells. *Journal of Nutritional Biochemistry*, *19*, 313–319.

Liao, Y. C., Shih, Y. W., Chao, C. H., Lee, X. Y., & Chiang, T. A. (2009). Involvement of the ERK signaling pathway in fisetin reduces invasion and migration in the human lung cancer cell line A549. *Journal of Agricultural and Food Chemistry*, *57*, 8933–8941.

Lim, T. G., Kwon, J. Y., Kim, J., Song, N. R., Lee, K. M., Heo, Y. S., Lee, H. J., & Lee, K. W. (2011). Cyanidin-3-glucoside suppresses B[*a*]PDE-induced cyclooxygenase-2 expression by directly inhibiting Fyn kinase activity. *Biochemical Pharmacology*, *82*, 167–174.

Lim, D. Y., & Park, J. H. (2009). Induction of p53 contributes to apoptosis of HCT-116 human colon cancer cells induced by the dietary compound fisetin. *American Journal of Physiology. Gastrointestinal and Liver Physiology*, *296*, G1060–G1068.

Liu, R. H. (2004). Potential synergy of phytochemicals in cancer prevention: Mechanism of action. *Journal of Nutrition*, *134*, S3479–S3485.

Londono, J. L., de Lima, V. R., Lara, O., Gil, A., Pasa, T. B. C., Arango, G. J., & Pineda, J. R. R. (2010). Clean recovery of antioxidant flavonoids from citrus peel: Optimizing an aqueous ultrasound-assisted extraction method. *Food Chemistry*, *119*, 81–87.

Lopez-Lazaro, M., Willmore, E., & Austin, C. A. (2007). Cells lacking DNA topoisomerase II beta are resistant to genistein. *Journal of Natural Products*, *70*, 763–767.

Maher, P., Dargusch, R., Ehren, J. L., Okada, S., Sharma, K., & Schubert, D. (2011). Fisetin lowers methylglyoxal dependent protein glycation and limits the complications of diabetes. *PLoS One*, 6, e21226.

Monte, J., Abreu, A. C., Borges, A., Simoes, L. C., & Simoes, M. (2014). Antimicrobial activity of selected phytochemicals against *Escherichia coli* and *Staphylococcus aureus* and their biofilms. *Pathogens*, *3*, 473–498.

Nahum, A., Hirsch, K., Danilenko, M., Watts, C. K., Prall, O. W., Levy, J., & Sharoni, Y. (2001). Lycopene inhibition of cell cycle progression in breast and endometrial cancer cells is associated with reduction in cyclin D levels and retention of p27(Kipl) in the cyclin e-cdk2 complexes. *Oncogene*, *20*, 3428–3436.

Nam, K. N., Park, Y. M., Jung, H. J., Lee, J. Y., Min, B. D., Park, S. U., Jung, W. S., Cho, K. H., Park, J. H., Kang, I., Hong, J. W., & Lee, E. H. (2010). Anti-inflammatory effects of crocin and crocetin in rat brain microglial cells. *European Journal of Pharmacology*, *648*, 110–116.

Narod, S. A., Risch, H., Moslehi, R., Dorum, A., Neuhausen, S., Olsson, H., Provencher, D., Radice, P., Evans, G., Bishop, S., Brunet, J. S., & Ponder, B. A. (1998). Oral contraceptives and the risk of hereditary ovarian cancer. *The New England Journal of Medicine*, *339*, 424–428.

Ning, G., Tianhua, L., Xin, Y., & He, P. (2009). Constituents in *Desmodium blandum* and their antitumor activity. *Chinese Traditional and Herbal Drugs, 40*, 852–856.

Nothlings, U., Murphy, S. P., Wilkens, L. R., Henderson, B. E., & Kolonel, L. N. (2007). Flavonols and pancreatic cancer risk: The multiethnic cohort study. *American Journal of Epidemiology, 166*, 924–931.

Okwu, D. E. (2005). Phytochemicals, vitamin and mineral contents of two Nigeria medicinal plants. *International Journal of Molecular Medicine and Advance Sciences, 1*, 375–381.

Oyagbemi, A. A., Saba, A. B., & Azeez, O. I. (2010). Molecular targets of [6]-gingerol: Its potential roles in cancer chemoprevention. *Biofactors, 36*, 169–178.

Park, Y. J., Wen, J., Bang, S., Park, S. W., & Song, S. Y. (2006). [6]-Gingerol induces cell cycle arrest and cell death of mutant p53-expressing pancreatic cancer cells. *Yonsei Medical Journal, 47*, 688–697.

Patil, P. S., & Shettigar, R. (2010). An advancement of analytical techniques in herbal research. *Journal of Advanced Scientific Research, 1*, 8–14.

Peter, M. (2013). Ethnobotanical study of some selected medicinal plants used by traditional healers in Limpopo province (South Africa). *American Journal of Research Communication, 1*, 8–23.

Rahman, A. U., & Choudhary, M. I. (2001). Bioactive natural products as potential source of new pharmacophores: A theory of memory. *Pure and Applied Chemistry, 73*, 555–560.

Raynal, N. J., Momparler, L., Charbonneau, M., & Momparler, R. L. (2008). Antileukemic activity of genistein, a major isoflavone present in soy products. *Journal of Natural Products, 71*, 3–7.

Rhazi, N., Hannache, H., Oumam, M., Sesbou, A., Charrier, B., Pizzi, A., & Charrier-El Bouhtoury, F. (2015). Green extraction process of tannins obtained from Moroccan *Acacia mollissima* barks by microwave: Modeling and optimization of the process using the response surface methodology RSM. *Arabian Journal of Chemistry, 12*, 2668–2684.

Rhode, J., Fogoros, S., Zick, S., Wahl, H., Griffith, K. A., Huang, J., & Liu, J. R. (2007). Ginger inhibits cell growth and modulates angiogenic factors in ovarian cancer cells. *BMC Complementary and Alternative Medicine, 7*, 44.

Sala, A., Recio, M. D., Giner, R. M., Manez, S., Tournier, H., Schinella, G., & Rios, J. L. (2002). Anti-inflammatory and antioxidant properties of *Helichrysum italicum*. *Journal of Pharmacy and Pharmacology, 54*, 365–371.

Samarghandian, S., Boskabady, M. H., & Davoodi, S. (2011). Use of *in vitro* assays to assess the potential antiproliferative and cytotoxic effects of saffron (*Crocus sativus* L.) in human lung cancer cell line. *Pharmacognosy Magazine, 6*, 309–314.

Samarghandian, S., TavakkolAfshari, J., & Davoodi, S. (2010). Suppression of pulmonary tumor promotion and induction of apoptosis by *Crocus sativus* L. extraction. *Applied Biochemistry and Biotechnology, 164*, 238–247.

Sanchez, Y., Amran, D., de Blas, E., & Aller, P. (2009). Regulation of genistein-induced differentiation in human acute myeloid leukaemia cells (HL60, NB4) protein kinase modulation and reactive oxygen species generation. *Biochemistry & Pharmacology, 77*, 384–396.

Shami, A. M. (2015). Isolation and identification of anthraquinones extracted from *Morinda citrifolia* L. (Rubiaceae). *Annals of Chromatography and Separation Techniques, 1*, 1012.

Shapiro, T. A., Fahey, J. W., Wade, K. L., Stephenson, K. K., & Talalay, P. (2001). Chemoprotective glucosinolates and isothiocyanates of broccoli sprouts: Metabolism and excretion in humans. *Cancer Epidemiology, Biomarkers & Prevention, 10*, 501–508.

Shirwaikar, A., Rajendran, K., & Punitha, I. S. (2006). *In vitro* anti-oxidant studies on the benzyl tetra isoquinoline alkaloid berberine. *Biological & Pharmaceutical Bulletin, 29*, 1906–1910.

Smith, R. D., Wright, B. W., & Yonker, C. R. (1988). Supercritical fluid chromatography: Current status and prognosis. *Analytical Chemistry, 60*, A1323–A1336.

Stray, F., & Storchova, H. (1991) *The natural guide to medicinal herbs and plants* (2nd ed., p. 223). Dorset House Publishing.

Wang, W., Bringe, N. A., Berhow, M. A., & Gonzalez de Mejia, E. (2008). Beta-conglycinins among sources of bioactives in hydrolysates of different soybean varieties that inhibit leukemia cells *in vitro*. *Journal of Agricultural and Food Chemistry, 56*, 4012–4020.

Watson, A. A., Fleet, G. W. J., Asano, N., Molyneux, R. J., & Nash, R. J. (2001). Polyhydroxylated alkaloids: Natural occurrence and therapeutic applications. *Phytochemistry, 56*, 265–295.

Williams, R. J., Spencer, J. P., & Rice-Evans, C. (2004). Flavonoids: Antioxidants or signalling molecules. *Free Radical Biology & Medicine, 36*, 838–849.

Wong, T. S., Roccatano, D., Zacharias, M., & Schwaneberg, U. (2006). A statistical analysis of random mutagenesis methods used for directed protein evolution. *Journal of Molecular Biology, 355*, 858–871.

Wood, J. G., Rogina, B., Lavu, S., Howitz, K., Helfand, S. L., Tatar, M., & Sinclair, D. (2004). Sirtuin activators mimic caloric restriction and delay ageing in metazoans. *Nature, 430*, 686–689.

Wrobleski, A., Sahasrabudhe, K., & Aube, J. (2004). Asymmetric total synthesis of dendrobatid alkaloids: Preparation of indolizidine 251F and its 3-desmethyl analogue using an intramolecular Schmidt reaction strategy. *Journal of the American Chemical Society, 28*, 5475–5481.

Xie, P., Chen, S., Liang, Y. Z., Wang, X., Tian, R., & Upton, R. (2006). Chromatographic fingerprint analysis: A rational approach for quality assessment of traditional Chinese herbal medicine. *Journal of Chromatography, 1112*, 171–180.

Xu, M., Bower, K. A., Wang, S., Frank, J. A., Chen, G., Ding, M., Wang, S., Shi, X., Ke, Z., & Luo, J. (2011) Cyanidin-3-glucoside inhibits ethanol-induced invasion of breast cancer cells overexpressing ErbB2. *Molecular Cancer, 9*, 285.

Yamasaki, M., Fujita, S., Ishiyama, E., Mukai, A., Madhyastha, H., Sakakibara, Y., Suiko, M., Hatakeyama, K., Nemoto, T., Morishita, K., Kataoka, H., Tsubouchi, H., & Nishiyama, K. (2007). Soy-derived isoflavones inhibit the growth of adult t-cell leukemia cells *in vitro* and *in vivo*. *Cancer Science, 98*, 1740–1746.

Yan, L. L., Zhang, Y. J., Gao, W. Y., Man, S. L., & Wang, Y. (2009). *In vitro* and *in vivo* anticancer activity of steroid saponins of *Paris polyphylla* var. *yunnanensis*. *Experimental Oncology, 31*, 27–32.

Zikri, N. N., Ried, K. M., Wang, L. S., Lechner, J., Schwartz, S. J., & Stoner, G. D. (2009). Black raspberry components inhibit proliferation, induce apoptosis, and modulate gene expression in rat esophageal epithelial cells. *Nutrition and Cancer, 61*, 816–826.

5 Herbal Formulations Used for Treatment/ Management of Cancer

Ashok Koshta, Neelesh Malviya, and Sapna Malviya

CONTENTS

5.1 INTRODUCTION

According to the literature, traditional medicine is "the knowledge, skills, and practices based on the theories, beliefs, and experiences unique to different cultures, employed in the maintenance of health and in the prevention, diagnosis, improvement, or treatment of physical and mental illness" (World Health Organization). The use of herbs is a crucial component of all traditional medical systems, and the focus is typically on the patient's general health rather than the specific illness or sickness they are experiencing.

Traditional Chinese medicine (TCM) is a crucial illustration of how collected, ancient knowledge is used in modern healthcare with a holistic approach. TCM has a longer than 3000-year history. The earliest known herbal document in the world is the Divine Farmer's Classic of Herbalism, which was put together in China some 2000 years ago. Numerous herbal pharmacopeias and monographs on specific herbs

DOI: 10.1201/9781003251712-5

have been established from the accumulated and meticulously gathered material on herbs.

Drugs created through chemical synthesis and their widespread manufacture over the past century have completely changed the way that most people access healthcare worldwide. Despite this, many populations in developing countries still rely on conventional physicians and herbal medicines for their basic care. In order to meet their medical needs, up to 90% of the people in Africa and 70% of the population in India rely on traditional medicine. In China, traditional medicine accounts for around 40% of all healthcare services, and traditional medicine departments are present in more than 90% of general hospitals (WHO 2005). The use of traditional medicine has spread outside underdeveloped countries over the past 20 years, as ethnobotanical has become more popular and interest in natural cures has increased there. In the United States, 12% of children and 38% of adults reported using traditional medicine in some manner in 2007.

The most frequent justifications for choosing traditional medicine are that it is more cost-effective, more in line with the patient's philosophy, concerns about the side effects of chemical (synthetic) medicines, satisfies a desire for more individualized healthcare, and makes health information more widely available to the general public. Herbal remedies are primarily used for chronic, as opposed to life-threatening, diseases and for health promotion. However, the use of traditional treatments rises when modern medicine fails to effectively treat an illness, as is the case with advanced cancer and emerging infectious diseases. Traditional medicines are also widely regarded as being natural, harmless, and non-toxic.

Whether or not people have physical or financial access to allopathic medicine, traditional medicine is a thriving global commercial industry that offers a crucial healthcare service. The cost of "alternative" therapy in the United States was estimated at US$13.7 billion in 1990. By 1997, it had doubled, with herbal medicines experiencing the fastest growth of any alternative therapy. Annual spending on conventional medicine is projected to reach US$80 million, US$1 billion, and US$2.3 billion in Australia, Canada, and the United Kingdom, respectively. These statistics reflect the widespread use of herbal and other traditional medicines in various healthcare systems.

Herbal medicine is widely utilized in India, where the business uses 960 plant species in total, 178 of which are employed in significant volumes that exceed 100 metric tonnes annually. In 1995, China produced herbal medicines worth a total of 17.6 billion Chinese yuan. Western Europe's annual revenues in 2003–2004 hit $5 billion due to the continuation of this trend. Sales of herbal products in China reached US$14 billion in 2005, and in Brazil, sales of herbal treatments brought in US$160 million in 2007. The anticipated yearly global market for these goods was close to US$60 billion.

Currently, a variety of diseases and disorders, including inflammation, cardiovascular disease, prostate problems, depression, and immune system support, to name but a few, are treated with herbs, encompassing both acute and chronic ailments. More than 70% of the 177 drugs used to treat cancer throughout the world are based on natural substances or their mimics, many of which have undergone combinatorial chemical enhancement. More than 60% of cancer treatments that are now available or being tested are based on natural ingredients. According to another estimate, around 25% of medications prescribed worldwide contain active compounds derived from plants; there are now 121 such substances in use. Between 2005 and 2007, 13 drugs made on natural ingredients were approved in the United States. Of the 252 drugs on the World Health Organization's list of essential pharmaceuticals, 11% are only produced from plants, and more than 100 natural product-based medications are now undergoing clinical studies.

5.2 MARKETED AVAILABLE PRODUCTS FOR CANCER TREATMENT/MANAGEMENT

5.2.1 Maha Herbals Karso Peace

Balancing the tridoshas and trigunas improves the body tissues' ability to heal naturally, aids in the recovery from chronic illnesses, mitigates the side effects of chemotherapy, and radiotherapy, and supports the body's ability to heal itself by eradicating free radical damage.

It is primarily given for disorders like carcinoma, TB, secondary metastasis, degenerative conditions, and immune deficiency conditions as one to two pills (500 mg each) taken twice daily or as directed by the doctor (Figure 5.1).

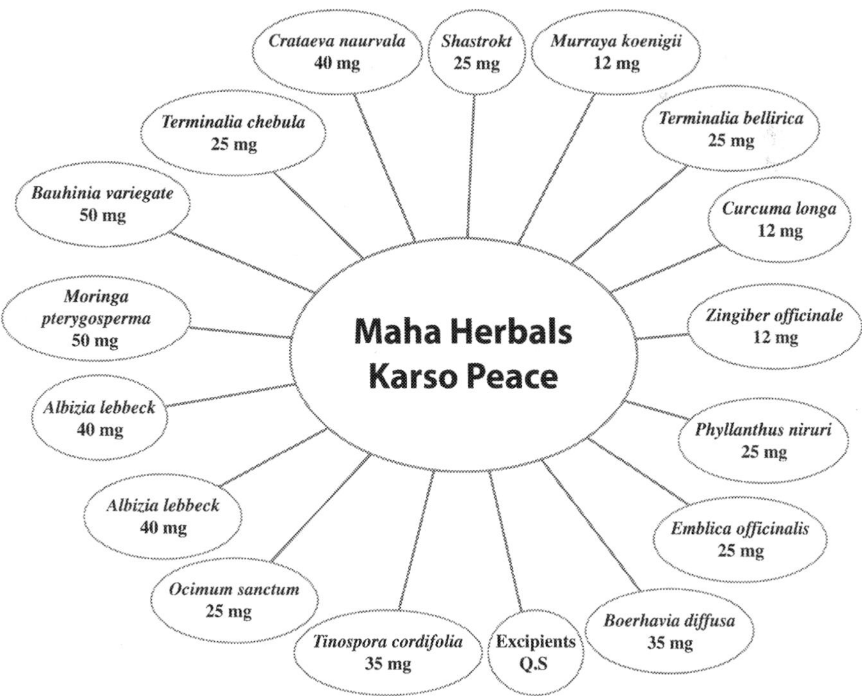

FIGURE 5.1 Maha Herbals Karso Peace.

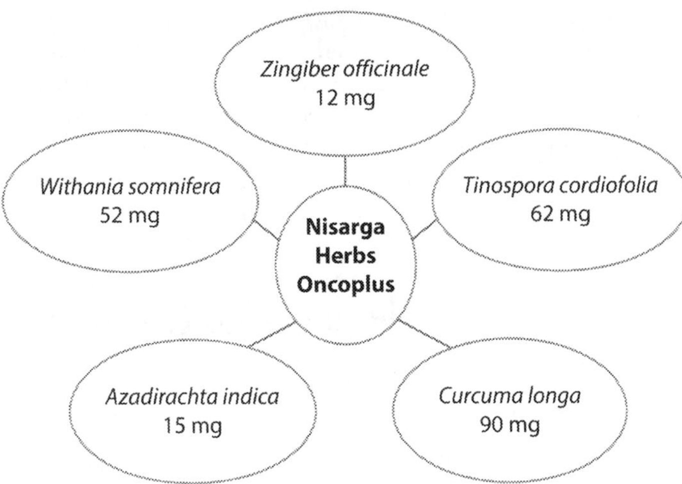

FIGURE 5.2 Nisarga Herbs Oncoplus.

5.2.2 Nisarga Herbs Oncoplus

Specifically developed for cancer patients undergoing treatment, Oncoplus is an antioxidant cellular support formula that boosts the immune system, restores the patient's mental clarity and vitality, and also aids in reducing nausea; it enhances liver function and the body's capacity to deal with inflammation. The capsule form of this remedy should be taken once or twice a day with water (Figure 5.2).

5.2.3 Oro-T Oral Rinse

For the prevention and treatment of oral mucositis (mouth ulcers) in cancer patients receiving radiotherapy and chemotherapy, the phytopharmaceutical formulation Oro-T Oral Rinse is advised (Figure 5.3). Additionally, it helps with the treatment

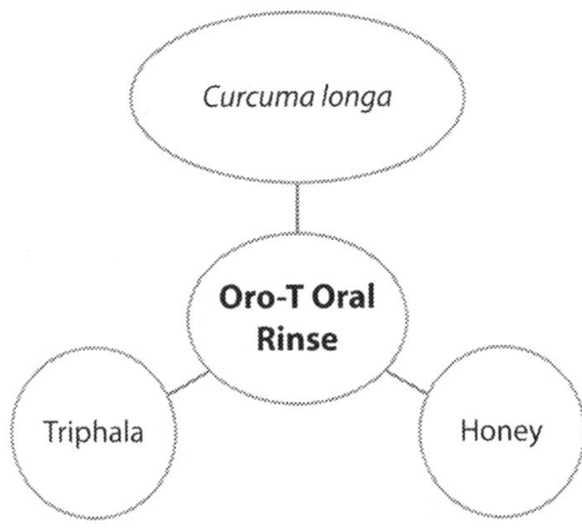

FIGURE 5.3 Oro-T Oral Rinse.

of oral lesions when the oral mucosa has thickened and made it difficult to open the mouth. Additionally, Oro-T Oral Rinse eases the discomfort and swelling associated with a sore throat (pharyngitis). It lessens pain, burning, inflammation, and the chance of subsequent infections. Oro-T Oral Rinse eventually encourages quicker recovery. Additionally, it aids in the treatment of oral submucous fibrosis (OSF or OSMF), a debilitating condition that causes mouth ulcers, a burning feeling in the mouth, and a thickening of the oral mucosa that makes it difficult to open the mouth. The majority of OSF or OSMF cases are related to long-term paan (betel) and gutka users.

It can be purchased as a monophasic formulation for gargling. Oral rinsing must be done four to six times a day for at least a minute. This mouthwash is intended to treat oral conditions like pharyngitis, oral submucous fibrosis, and oral mucositis.

5.2.4 Baidyanath Kanchanar Guggulu

In addition to balancing the kapha dosha, Baidyanath Kanchanar Guggulu also helps maintain a healthy lymphatic system. Facilitating the removal of inflammatory poisons, it reduces the buildup of Kapha in the tissues. In addition, it is an anti-inflammatory tonic (Figure 5.4). It is given for glands, malignant ulcers, syphilis, fistula, scrofula, and other conditions. As already mentioned, a classic Ayurvedic remedy for kapha

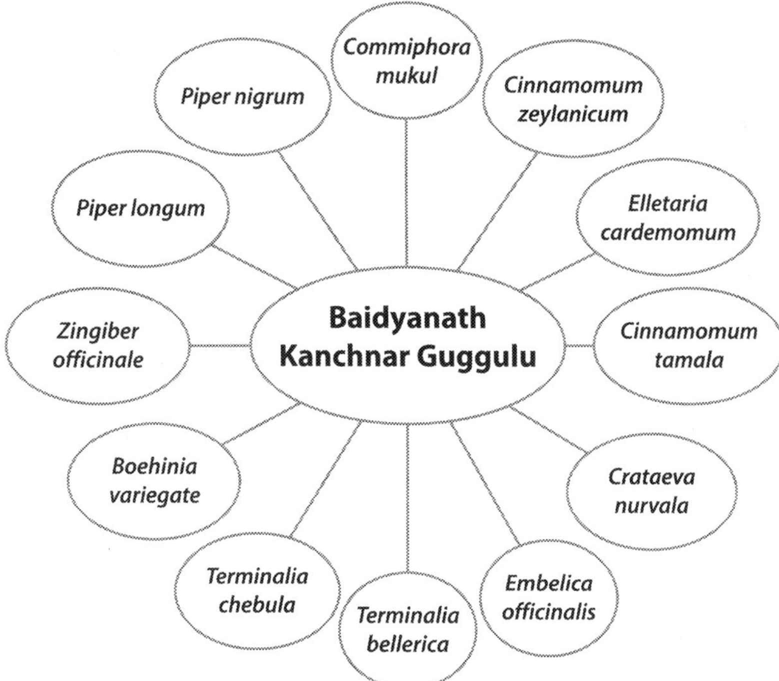

FIGURE 5.4 Baidyanath Kanchanar Guggulu.

accumulations in the tissues is called Kanchanar Guggulu. Deeper within the body, kapha might appear as enlarged lymph nodes, cysts, or growths.

Guggulu is combined with potent decongestants like kanchanar, triphala, and trikatu to dissolve and eradicate hardened kapha, gandamala, cysts, oedema, lymphatic congestion, sinus congestion, benign tumours, or rasoli or ganth. This cleansing mixture helps the lymphatic and digestive systems' healthy operation, assisting in the reduction of additional kapha imbalances. Two to four tablets should be taken with warm water before meals. It is prescribed for people with hypothyroidism, polycystic ovary syndrome (PCOS), lipomas (fatty tissue tumours), obesity or being overweight, cancer, tumours, cysts (including polycystic kidney disease and liver cysts), goitres, wounds, boils, fistulas, and skin conditions.

5.2.5 Shakti Drops

An efficient salutatory drug of Sri Sri Tattva for immune modulator activity is Shakti drops, an ayurvedic poly-herbal preparation (Figure 5.5). It is an extract made entirely of pure water from eight organic herbs. A person's general immunity is boosted by the formulation, which is made in a lab using real organic plant medicines. The standardization process is carried out in accordance with Indian Ayurvedic formulary guidelines and is based on physiochemical parameters and antibacterial investigations.

The Shakti drops' powerful antioxidant and free radical-scavenging action, according to the study's antioxidant activity results, justifies their usage in Rasayana.

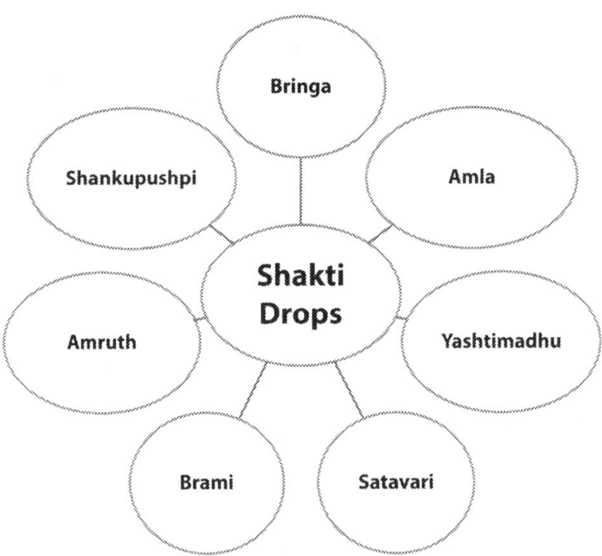

FIGURE 5.5 Shakti Drops.

5.2.6 TrueBasics Lung Detox (For Lung Cancer Management)

It defends, sanitizes, and fortifies the lungs. Complete lung care is provided by this product, which is made with powerful herbal extracts and clinically proven components, including BioPerine and ginger. It is available in a tablet form that should be taken once before breakfast.

Quercetin, which helps minimize airway hyperreactivity and limits excessive mucus formation, is added to the product. Additionally, it contains licorice, which is well known for clearing the airways by accelerating mucus evacuation. This anti-inflammatory mixture, which contains *Curcuma longa*, *Nelumbo nucifera*, and *Echinacea*, aids in reducing lung damage.

5.2.6.1 Benefits

1. Licorice and quercetin, two ingredients in TrueBasics Lung Detox supplements (Figure 5.6), limit the overproduction of mucus and encourage its expulsion, which helps to clear the respiratory tract.
2. Dyspnoea relief—*Echinacea* extracts may have bronchodilatory properties that help with breathlessness and airway constriction. Magnesium's

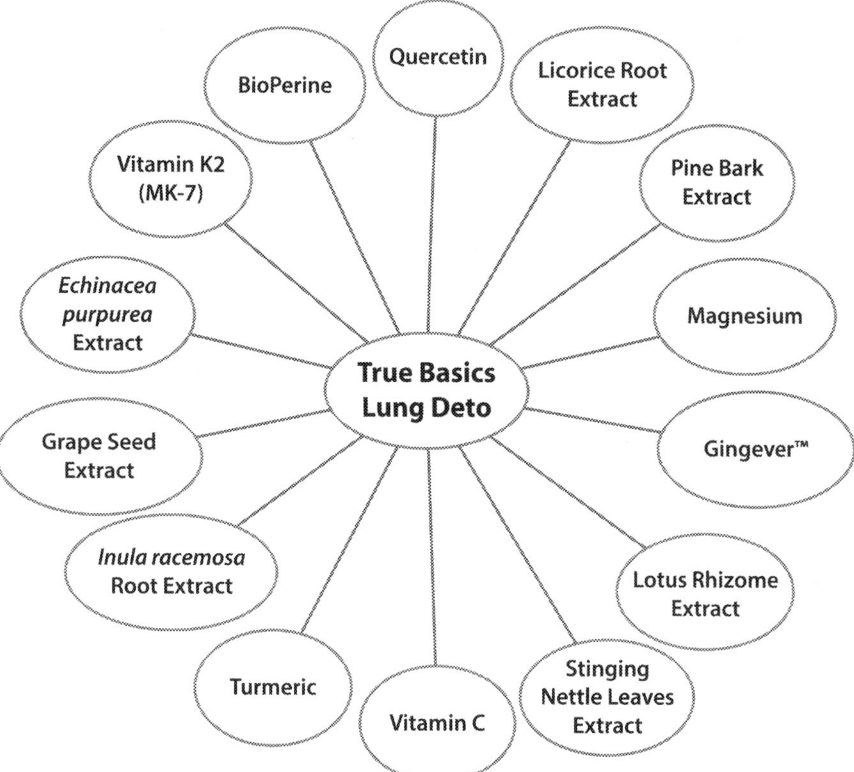

FIGURE 5.6 TrueBasics Lung Detox.

bronchodilators and anti-inflammatory properties help to reduce lung congestion.
3. Containing flavonoids from pine bark extract, which are strong antioxidants that can shield lung tissues from the damaging effects of free radicals, this substance offers protection from pollutants. Due to their potent ability to scavenge free radicals, the polyphenols in the grape seed extract in this product may be able to reduce the oxidative stress brought on by pollution.
4. Immune support—by promoting the proliferation of immune cells like B and T cells, vitamin C, turmeric, and *Echinacea* may help boost immunity.

5.2.7 Papaya Complete

A powerful nutritional supplement called Pure Nutrition Papaya Complete is created from papaya leaf and fruit extracts. Papain and chymopapain, which aid in the breakdown of proteins to naturally speed up metabolism, are abundant in this food. One of the best and richest sources of antioxidants is papaya leaf extract. The supplement boosts the immune system, promotes digestion, and helps maintain platelet count. It is recommended to take one pill after each meal, twice a day.

Papaya leaves are rich in many phytochemicals, including alkaloids, saponins, and tannins. The bone marrow is protected by these alkaloids, which stop the blood's platelet oxidation. Papaya is known for its regenerating and purifying qualities. Antioxidants included in papaya leaf extract help fight off and prevent illnesses, hence promoting excellent health.

Papaya contains large amounts of proteolytic enzymes like papain, chymopapain, caricain, and glycyl endopeptidase. The digestive processes are aided by these enzymes. Papaya leaf and papaya leaf extract are excellent for treating digestive problems and for calming any gastrointestinal disturbances. A potent enzyme called papain aids in the breakdown and digestion of protein. In turn, this relieves indigestion and stomach discomfort.

5.2.7.1 Benefits

- Spectacular papaya fruit and leaf extract combo (Figure 5.7)
- Aids in boosting platelet count
- Boosts resistance
- Promotes healthy bowel movement and aids with digestion

5.2.8 Noni Gold Liquid

The concentrated noni juice Pure Nutrition Noni Gold is enhanced with scientifically studied nutrients, including Garcinia, Aloe vera, Amla, Ashwagandha, Grape Seed extract, and others. Fresh noni plants, sometimes referred to as Indian Mulberry, are used to make Pure Nutrition Noni Gold. Numerous vitamins, minerals, amino acids, and other phytonutrients included in noni support cell and body renewal while preserving general health. Noni Gold's powerful antioxidants, including beta-carotene, iridoids, and vitamins C and E, can stop free radicals from causing cellular harm.

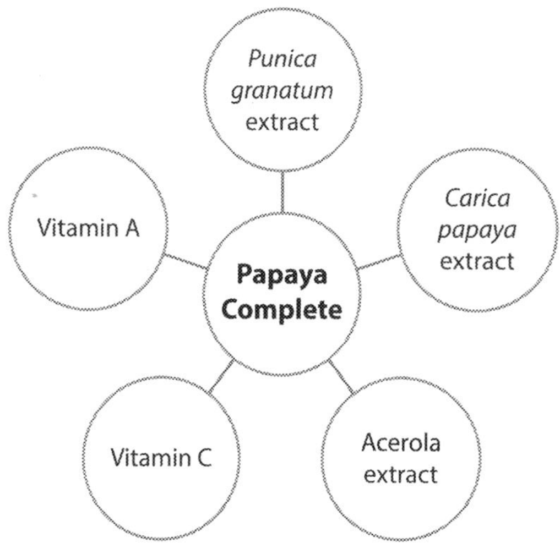

FIGURE 5.7 Papaya complete.

Antioxidants present in noni juice may help reduce the chance of developing long-term illnesses, including diabetes and heart disease. In addition to boosting immunity, Pure Nutrition Noni Gold also lessens arthritis pain (Figure 5.8), which is brought on by increased inflammation, and oxidative stress by its antibacterial, antifungal, anti-inflammatory, and anti-histamine characteristics. Being a potent source

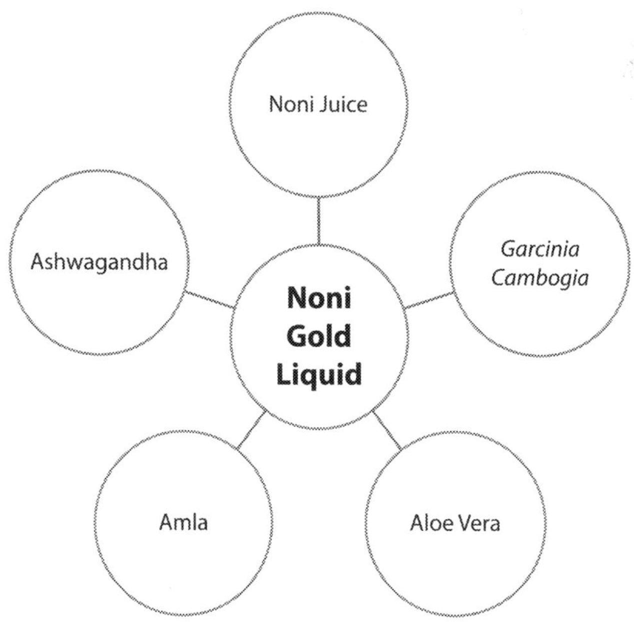

FIGURE 5.8 Noni Gold Liquid.

of antioxidants, it aids in the body's detoxification process, enhances skin health, and fights infections, the common cold, and the flu. It increases energy levels, fights general bodily weakness, and enhances the body's overall physical performance. Additionally, it facilitates weight loss and enhances digestion. The usual dose is 15–30 ml of juice in water, taken first thing in the morning.

5.2.8.1 Benefits

- Contributes to increased energy and enhanced physical performance
- Promotes weight loss and helps with digestion
- Enhances skin health, detoxification, and infection prevention
- Assists in reducing pain and swelling associated with arthritis

5.2.9 Dr Thankis Tum Can Capsule

This aids in the treatment of different types of tumours and cancers, abscesses in any part of the human body, multiple sclerosis, brain tumours, brain cancer, Ebola, HIV/AIDS, hepatitis b and c, swine flu, *Escherichia coli*, neuralgia, unidentified harmful viruses, germs, bacteria, Naru Bala, urinary retention, etc. (Figure 5.9).

Key Advantages:

- It functions as a therapeutic therapy without producing any negative side effects. Contrarily, chemotherapy based on chemicals causes common side effects such as nausea, a propensity to vomit, hair loss, weight loss, etc.
- It protects the immune system and wards off dangerous illnesses, it is therefore incredibly advantageous to people.

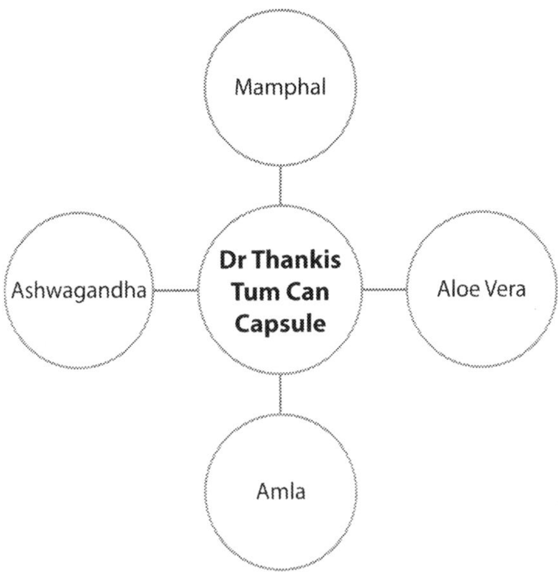

FIGURE 5.9 Dr Thankis Tum Can Capsule.

- This medication enables the patient feel energized and healthier.
- This medication can also be used to treat liver, cardiac, and asthma conditions.

5.2.10 Kudos CM9

The Ministry of Science & Technology's CSIR–IICB (Council of Scientific & Industrial Research–Indian Institute of Chemical Biology) has recently announced its first-ever Ayurvedic invention for cancer treatment—Kudos CM9. This is the first non-chemical, non-synthetic Ayurvedic drug ever developed to fight the deadly cancer threat.

Both benign and malignant tumours can be treated with this medicine. It is extremely helpful in treating early-stage/early-diagnosed cancer patients and in preventing cancer relapse in patients having chemotherapy, radiation, and hormone therapy. It also controls the further spread of cancer growth in patients undergoing these treatments.

It safeguards healthy cells while solely killing malignant cells. As a result, increasing the patients' immunological capacity is a well-known immunomodulator and a gold shield for providing immunotherapy to cancer patients. It aids in the battle against cancer by obstructing particular enzymes and hormones that cause cancer, those at high risk can be avoided; it has antimutagenic, chemoprotective, and radioprotective qualities as well.

It combats and eradicates cancer at every step, including its genesis, growth, and spread. Patients who have developed a resistance to hormone therapy, radiation, or chemotherapy can still benefit from it. It is also very helpful for patients who cannot undergo surgery, chemotherapy, or radiotherapy because of advanced age, pronounced frailty, or any other reason.

5.3 CONCLUSION

In most human civilizations, different plants have been utilized for centuries or even millennia for both cure and prevention of illness. This chapter serves as an example of how, over the past 20 years, the renewed public interest and research efforts from the scientific and medical communities around the world have produced a significant amount of data, including clinical studies and trials on the pharmacological effects, use, and development into future medicines of herbs and derivative medicinal phytochemicals as anti-tumour and chemoprevention agents. Even if many conventional treatments or combinations of herbs have undergone verification and improvement, systematic standardized research, the application of FDA regulatory processes, and clearly defined clinical trials are still fairly restricted and need to be aggressively pursued. To hasten the discovery and development of new phytomedicines and botanical pharmaceuticals, traditional medical practices around the world have amassed a variety of records and documentation.

BIBLIOGRAPHY

Balsano, C., & Alisi, A. (2009). Antioxidant effects of natural bioactive compounds. *Current Pharmaceutical Design*, *15*, 3063–3073.

Barnes, P. M., Bloom, B., & Nahin, R. (2008). *Complementary and alternative medicine use among adults and children: United States, 2007* (CDC National Health Statistics Report #12). Available from www.cdc.gov/nchs/data/nhsr/nhsr012.pdf (accessed 5 Nov).

Beckman, K. B., & Ames, B. (1998). The free radical theory of ageing matures. *Physiological Reviews*, *78*, 47–81.

Benzie, I. F. F., & Wachtel-Galor, S. (2009). Biomarkers in long-term vegetarian diets. *Advances in Clinical Chemistry*, *47*, 170–208.

Benzie, I. F. F., & Wachtel-Galor, S. (2010). Vegetarian diets and public health: Biomarker and redox connections. *Antioxidants & Redox Signaling*, *13*(10), 1575–1591.

Bozzetti, F. (2003). Nutritional issues in the care of the elderly patient. *Critical Reviews in Oncology/Hematology*, *48*, 113–121.

Brower, V. (2008). Back to nature: Extinction of medicinal plants threatens drug discovery. *Journal of the National Cancer Institute*, *100*, 838–839.

https://mahaherbals.biz/products/special-preparations/karso-peace

https://www.clickoncare.com/nisarga-herbs-oncoplus-60-capsules

https://himalayawellness.in/products/oro-t-oral-rinse

https://organiclab.pl/en/cancer-care/182-baidyanath-kanchnar-guggulu-supports-actively-immunological-and-lymphatic-systems.html

https://www.srisritattva.com/products/shop-shakti-drops-immunity-booster

https://www.truebasics.com/sv/truebasics-lung-detox/SP-61159?navKey

https://purenutrition.in/products/papaya-complete-supports-platelet-immunity-digestion-60-veg-capsules

https://purenutrition.in/products/noni-gold-liquid-for-healthy-life-style-

https://www.vishlaagrotech.com/product-details/dr-thankis-tum-can-capsule-16.html

https://www.biospectrumindia.com/news/58/12739/ministry-of-st-csir-iicb-release-first-ayurvedic-cancer-drug.html

Tang, S. Y., & Halliwell, B. (2010. Medicinal plants and antioxidants: What do we learn from cell culture and *Caenorhabditis elegans* studies? *Biochemical and Biophysical Research Communications*, *394*, 1–5.

Tilburt, J. C., & Kaptchuk, T. J. (2008). Herbal medicine research and global health: An ethical analysis. *Bulletin of the World Health Organization*, *86*(8), 594–599.

U.S. Government Accountability Office (GAO). (2010). *Herbal dietary supplements: Examples of deceptive or questionable marketing practices and potentially dangerous advice* (GAO-10-662T).

6 Alternative Approaches for Treatment of Cancer

Ankur Joshi, Neelesh Malviya, and Sapna Malviya

CONTENTS

6.1 INTRODUCTION

One of the leading causes of death in the world is cancer, and during the past 10 years, numerous research projects have concentrated on developing novel treatments to lessen the adverse effects of existing ones. As cancer progresses, tumours become extremely heterogeneous, resulting in a mixed population of cells with a range of molecular characteristics and therapeutic responses. This variability, which is crucial to the establishment of resistance phenotypes encouraged by a selection pressure upon treatment delivery, can be observed at both the geographical and temporal levels. Typically, cancer is handled as a single, uniform disease, and tumours are seen as an entire cell population. Therefore, it is essential to have a thorough understanding of these complicated events in order to develop effective and precise treatments. In order to administer traditional chemotherapeutic medications *in vivo*, increase their bioavailability and concentration around tumour tissues, and enhance their release profile, nanomedicine provides a flexible platform of biocompatible and biodegradable technologies. Nanoparticles can be used in a variety of processes, from therapy to medical diagnosis. Extracellular vesicles (EVs), which are involved in the formation of cancer, the change of the microenvironment, and the spread of

DOI: 10.1201/9781003251712-6

metastatic disease, have recently attracted a lot of attention as effective drug delivery systems.

Due to their antiproliferative and proapoptotic qualities, natural antioxidants and several phytochemicals have lately been offered as anti-cancer adjuvant therapy.

Another type of cancer treatment is targeted therapy, which focuses on one specific area, such as the intracellular organelles or the vasculature of a tumour, while sparing the surrounding tissue. This significantly improves the treatment's specificity while lowering its disadvantages.

Using targeted silencing mediated by small interfering RNAs (siRNAs), gene therapy, expression of genes inducing apoptosis, and wild-type tumour suppressors are two more potential opportunities that are currently being tested in several clinical trials across the globe.

By allowing for the localization of treatment in extremely small and precise locations, thermal ablation of malignancies and magnetic hyperthermia are presenting new potential for precision medicine. These techniques might serve as an alternative to more intrusive procedures like surgery.

Additionally, emerging sciences like radiomics and pathomics are assisting in the creation of innovative methods for gathering large amounts of data, developing novel therapeutic approaches, and accurately predicting patient responses, clinical outcomes, and cancer recurrence.

A general summary of the most cutting-edge and innovative cancer medicines was given in this narrative unit. There are also several tactics for cancer detection and treatment, their status in the clinical setting, underscoring their significance as novel anti-cancer approaches, and new strategies that are presently being researched and should outweigh the limitations of conventional therapies.

6.2 CANCER TREATMENT MODALITIES

By categorizing cancer therapy options into conventional (traditional) and advanced or new or modern categories, we may see how they differ from one another. Over half of all currently running medical treatment trials globally focus on cancer treatments today. The progression of the treatment depends on factors like the type of cancer, where it is located, and how severe it is. Surgery, chemotherapy, and radiotherapy are the most often used conventional treatment modalities. Modern modalities include hormone therapy, anti-angiogenic, stem cell treatments, immunotherapy, and immunotherapy based on dendritic cells.

6.3 CONVENTIONAL CANCER THERAPIES

The standard cancer treatment approaches that are most frequently advised involve surgically removing the tumours, followed by radiotherapy using X-rays and/or chemotherapy. Surgery is one of these treatments that works best while the disease is still in its early stages. Radiation therapy has the potential to harm healthy tissues, cells, and organs. Despite the fact that chemotherapy has decreased morbidity and death, almost all chemotherapeutic drugs harm healthy cells, particularly those that divide and expand quickly. A significant issue with chemotherapy is drug resistance, which occurs when cancer cells that were originally inhibited by an anti-cancer treatment start to become resistant to the agent. Reduced drug absorption and increased drug efflux are the main contributors to this limitations of traditional chemotherapy, including difficult dosage selection, lack of selectivity, quick drug metabolism, and mostly negative side effects.

6.4 ADVANCED AND INNOVATIVE CANCER THERAPIES

Drug resistance and its delivery mechanisms are the biggest barriers to treating cancer and reducing its symptoms, yet there are presently several authorized therapy modalities and medications. Due to aberrant blood artery architecture and tumour biology, traditional cancer is less effective than it formerly was. The most cutting-edge and creative cancer therapeutic approaches are listed below, along with their advantages and disadvantages.

6.4.1 Stem Cells Therapy

In the bone marrow (BM), stem cells are undifferentiated cells with the capacity to develop into any kind of body cell. Another cancer treatment method that is thought to be both safe and efficient is the use of stem cells. The use of stem cells is yet in the exploratory stage of clinical trials; one potential application is the regeneration of other damaged tissue. BM, fat, and connective tissue–derived mesenchymal stem cells (MSCs) are now employed in clinical trials.

6.4.1.1 Pluripotent Stem Cells

The embryo's homogeneous inner mass cells, known as embryonic stem cells (ESCs), can give rise to any type of cell, with the exception of those found inside the placenta. A breakthrough in cell biology occurred in 2006 with the development of Yamanaka factors, which allowed physical cells in a culture to become induced pluripotent stem cells (iPSCs). Because iPSCs and ESCs have the same traits, there are no ethical concerns associated with embryo destruction. For the development of anti-cancer vaccines as well as the generation of effector T cells and natural killer (NK) cells, hematopoietic embryonic stem cells (hESCs) and iPSCs are now utilized.

6.4.1.2 Adult Stem Cells

Hematopoietic stem cells (HSCs), mesenchymal stem cells (MSCs), and neural stem cells (NSCs) are adult stem cell categories that are frequently used in tumour therapy. All adult blood cells in the body can be formed by HSCs, which are found in BM. Only the infusion of HSCs produced from cord blood is currently authorized by the Food and Drug Administration (FDA) to treat multiple myeloma and leukaemia. MSCs are present in a variety of tissues and organs and are crucial for tissue regeneration into osteocytes, adipocytes, and chondrocytes as well as for tissue repair.

MSCs are employed in conjunction with other methods of treating malignancies due to their unique biological properties. NSCs are employed to treat both primary and metastatic breast and other malignancies since they can self-renew and produce new neurones and glial cells.

6.4.1.3 Cancer Stem Cells

Epigenetic alterations cause normal stem cells, precursor/progenitor cells, or cancer stem cells (CSCs) to develop. They play a part in the growth, metastasis, and recurrence of cancer, which suggests that they may be effective in treating solid tumours. There are various ways that stem cells can fight the tumour. One mechanism involves the HSCs quickly migrating into specific stem cell niches in the BM, after which the transplants go through the engraftment phase before producing specialized blood cells. The production of the matrix-degradable enzyme matrix metalloproteinase (MMP-2/9), interaction with endothelial cells through lymphocyte function–associated antigen (LFA-1), very late antigen (VLA-4/5), and CD44, and the active interaction between stem cell CXCR4 receptors are all necessary for this pathway. The second mechanism is the tumour-tropic effect, in which MSCs migrate towards the tumour microenvironment (TM) after being drawn there by the tumour cells' secretions of CXCL16, SDF-1, CCL-25, and IL-6, and then differentiate within the tumour cells to aid in the growth of the tumour stroma. In addition to secreting paracrine factors including EVs and soluble substances, stem cells also have the ability to differentiate into all different types of blood cells through transplanted HSCs.

Cancer is typically treated with stem cell therapy employing a variety of techniques, such as HSC transplantation, MSC infusion, therapeutic carriers, the creation of immunological effector cells, and vaccine development. The following adverse effects were encountered with the stem cell cancer therapy strategy.

Tumourigenesis, unfavourable outcomes from allogeneic HSC transplantation, medication toxicity and treatment resistance, heightened immunological responses and autoimmunity, and viral infection are among the possible side effects. Despite these achievements, there are still problems that need to be looked into and resolved in the future, including therapeutic dose management, low cell targeting, and retention in tumour areas. Additionally, while early results from the use of stem cell therapies to treat tumours are very promising, more work needs to be done to increase their safety and effectiveness before they can be used in clinical trials. The licenced list of stem cell therapies is summarized in Table 6.1.

TABLE 6.1
Licensed Stem Cell Therapies

S. No.	Stem Cell Therapies	Examples	Authority	Indication
1	Pluripotent stem cells	iPSC (sipuleucel-T)	FDA	Prostate cancer
2	Adult stem cells	MSC-INFβ	FDA	Ovarian tumour
3	Cancer stem cells	Venetoclax	FDA	AML

6.4.2 Targeted Drug Therapy

Drugs or other chemicals that are "molecularly targeted," "molecularly targeted therapies," or "precision medicines" are considered targeted cancer therapy. These medications work by interfering with growth molecules, which prevents cancer from developing and relocating. The TM of an atypical tumour, which is made up of endothelial cells, pericytes, smooth muscle cells, fibroblasts, different inflammatory cells, dendritic cells, and CSCs, controls the genesis and growth of the tumour. The TM-forming cells actively engage with the malignant cells through a variety of signalling pathways and processes that are suitable for supporting a moderately high level of cellular growth. Therefore, employing TM circumstances to mediate efficient targeting strategies for cancer therapy is the field of study focus.

It is challenging to selectively treat cancer cells with traditional chemotherapy because they resemble normal cells. Cellular mechanisms, such as cell cycle arrest, induction of apoptosis, suppression of proliferation, and interference with metabolic reprogramming by targeted pharmacological therapy agents, intervene in order to address these issues. Two tactics that can be employed for the treatment of cancer include altering TM and targeting TM for drug delivery. Drugs used in targeted therapy do differ from those used in traditional chemotherapy in the manner they attack cancer cells while causing less harm to healthy cells, which is the programming that distinguishes cancer cells from healthy, normal cells.

The addition of erlotinib to regular chemotherapy boosted the survival rate for some diseases, bringing it from 17% to 24% in patients with advanced pancreatic cancer. Rituximab, sunitinib, and trastuzumab have all altered the treatment of renal cell carcinoma and breast cancer, respectively. Imatinib has had a significant impact on chronic myeloid leukaemia.

Based on how they operate or where they target, we may categorize the agents that target cells. Some enzymes act as growth signals for cancer cells. Some targeted medicines block the growth-stimulating enzymes that cancer cells use as signals. Enzyme inhibitors are the name of these medicines. By suppressing these cell signals, cancer can be prevented from developing and spreading.

Because they directly target the cell components that determine whether cells live or die, some targeted therapies are known as apoptosis-inducing medications. Examples include protein kinase B (PKB/Akt), which increases cell survival, and inhibitors of this enzyme are now in the preclinical stage of development.

These substances prevent tumours from forming new blood vessels, which helps to cut off the tumours' supply of blood and prevent tumour growth. Additionally, they halt the growth of tumours by reducing the amount of blood that reaches the tumour by blocking the activity of angiogenic factors like vascular endothelial growth factor (VEGF) or its receptors. According to the study, individuals with advanced colorectal cancer had their lives prolonged by months when Avastin (bevacizumab) was combined with chemotherapy that used the drug 5-fluorouracil.

6.4.2.1 Types of Target Agents

6.4.2.1.1 Monoclonal Antibodies

Intravenously given antibodies are synthetic immune system proteins that target specific targets on cancer cells. They have a higher percentage of human than murine components. Their attack strategies involve inducing the host immune system to attack the target cell, attaching to ligands or receptors to stop vital cancer cell operations, and delivering a deadly payload to the target cell, such as a radioisotope or poison. By conjugating with calicheamicin, the monoclonal antibody Gemtuzumab, for instance, targets CD-33 and is now utilized to treat AML. Ibritumomab tiuxetan, a 90Y metal isotope-based anti-CD20, was also developed for use in clinical therapy. Target agents of monoclonal antibodies are also used to deliver active medicines, prodrug activation enzymes, and chemotherapy toxins.

6.4.2.1.2 Small-molecule Inhibitors

These proteins have a reduced molecular weight (500 Da) than monoclonal antibodies, allowing them to easily cross plasma membranes and be ingested. Their primary function is to disrupt cellular processes by interfering with tyrosine kinase signalling that occurs intracellularly. This causes tyrosine kinase signalling to be inhibited, which sets off a molecular chain reaction that can stop cell growth, proliferation, migration, and angiogenesis in malignant tissues. The two examples of small-molecule inhibitors, 48 Gefitinib and erlotinib, block the kinase and EGFR, respectively, in patients with non-small cell lung cancer (NSCLC). Lapatinib and sorafenib are other medications that block EGFR/Erb-B2 receptor tyrosine kinase 2 (ERBB2) in order to treat ERBB2-positive breast cancer and VEGFR kinase in order to treat renal cancer, respectively.

6.4.2.1.3 Ablation Cancer Therapy

Ablation is a therapy method that eliminates tumours without removing them. It is typically recommended for small tumours that are less than 3 cm in size when surgery is not an option. For bigger tumours, embolization and ablation are combined. Due to the destruction of some of the normal tissue surrounding the tumour, this method may not be recommended for treating tumours close to major blood arteries, the diaphragm, or major bile ducts.

6.4.2.1.4 Thermal Ablation

This technique uses strong heat or hypothermia to target a focused zone inside and around the tumour for destruction. Similar to surgery, thermal ablation destroys the tumour and a margin of tissue that is 5–10 mm thick and looks to be normal but is really killed *in situ* before being absorbed by the body. The procedure is typically provided by a percutaneous or non-invasive technique and is analogous to open, laparoscopic, or endoscopic surgery. The type of tumour, its location, the doctor's preference, and the patient's health all play a role in the treatment chosen. Currently, clinical settings use cryoablation, high-intensity focused ultrasound, radio frequency ablation (RFA), and microwave ablation. With the help of a freeze-thaw procedure, cryoablation uses a hypothermic modality to cause tissue damage in others.

Except for cryoablation, all of these treatments work on the basis of hyperthermia. Cryoablation has the greatest potential to produce a post-ablative immunogenic response of any ablation procedure.

Recent research has shown that RFA and cryoablation, which are used as treatments for TM and in systemic circulation, can affect the immune system in addition to causing tissue disturbance. Ablation methods have been demonstrated to affect carcinogenesis because of the local inflammatory response that creates an immunogenic gene signature.

The benefit of this method over surgery is that it offers a minimally invasive (e.g., laparoscopically or percutaneously) or non-invasive approach to cancer therapy and attracts attention as a substitute for conventional surgical therapies.

6.4.2.1.5 Cryoablation

Cryoablation is one of the ablation methods that destroys large amounts of tissue by freezing it to fatal levels, followed by liquid formation. The majority of original tumours treated with this therapy are both benign and malignant. After experimenting with the use of cold temperatures by salt and ice solutions for the formation of local numbness before surgical operations in the nineteenth century, James Arnott found that the freezing temperatures can affect cancer cell viability. He recommended cryoablation as an appealing therapeutic choice that improved a patient's chance of survival.

The basis for cryoablation techniques is the Joule–Thomson phenomenon, which was extensively researched in the 1930s. It was found that utilizing liquid CO_2 at high pressure, liquid air, and liquid oxygen could produce ice crystals, which could then be used to cure lesions, warts, and keratosis. However, Allington took the position of liquid N_2 for the treatment of many skin lesion conditions after 1950.

6.4.2.1.6 RFA Therapy RFA is a minimally invasive method that uses high-frequency electrical currents to create a hyperthermic environment to kill cancer cells. Needle electrodes are guided into a tumour cell using imaging techniques, including ultrasound, computed tomography, or magnetic resonance imaging (MRI). RFA is typically the best method for treating small size tumours with a diameter of less than 3 cm. RFA may be combined with other traditional cancer therapy modalities. RFA can treat medium tumours after deploying deployable devices or numerous electrode systems (up to 5 cm diameter).

6.5 GENE THERAPY

In order to treat a particular condition, a defective gene is replaced with a healthy copy in a process known as gene therapy. The adenosine deaminase (ADA) gene was originally introduced to T cells in patients with severe combined immunodeficiency in 1990 using a retroviral vector (SCID). Two-thirds of over 2,900 ongoing clinical trials for gene therapy are focused on cancer. For cancer gene therapy, methods including the production of proapoptotic and chemosensitizing genes, wild-type tumour suppressor genes, genes able to elicit particular anti-tumour immune responses, and targeted silencing of oncogenes are being considered.

For the administration of the prodrug ganciclovir to stimulate its expression and induce particular cytotoxicity, thymidine kinase (TK) gene delivery is efficient.

Recently, the tumor suppressor gene p53, ears, the tumor suppressor gene p53, which is carried by vectors, has been evaluated for clinical use. A good response rate was demonstrated in NSCLC patients taking this drug alone and with chemotherapy. When combined with radiotherapy, Gendicine, a recombinant adenovirus carrying wild-type p53, caused full disease regression in patients with head and neck squamous cell carcinoma.

The correct circumstances and the finest delivery method to use are two issues that have been encountered with gene therapy. The therapy's genomic integration, limited effectiveness in certain patient subgroups, and significant risk of immune system neutralization have all been identified as downsides. The effective method of RNA interference (RNAi), which can result in targeted gene silence, has been applied in basic research and medicinal translation. By cleaving messenger RNA (mRNA) and interfering with protein synthesis, RNA-induced silencing complex (RISC) mediates the targeted gene silencing process. It is possible to create siRNAs to block specific targets, such as cell proliferation and metastatic invasion; as a result, particular molecular mechanisms are a catalyst for tumour development. This approach depends on siRNA-mediated gene suppression of cancer-related genes, transcription factors (such as the c-myc gene), and anti-apoptotic proteins (i.e. Kirsten rat sarcoma [KRAS]).

Safety, excellent efficacy, specificity, few adverse effects, and inexpensive production costs are benefits of siRNA-based medications. But occasionally, they can cause unintended consequences or trigger innate immune reactions, which are followed by a particular inflammation. The most effective way to get naked siRNAs to spontaneously cross cell membranes is chemical modification (insertion of a phosphorothioate at the 3′-end, addition of a 2′-*O*-methyl group, and modification by 2,4-dinitrophenol), lipid encapsulation, or conjugation with organic molecules (polymers, peptides, lipids, antibodies, small molecules). Sixty-nine simple electrostatic interactions between negatively charged nucleic acids and cationic liposomes make transfection simple and effective. They can be made up of *N*-[1-(2,3-dioleoyloxy) propyl]-*N*, *N*-trimethylammonium methyl sulphate and 1,2-dioleoyl-3-trimethylammonium propane (DOTAP) and *N*-[1-(2, 3-dioleyloxy)propyl]-*N*,*N*,*N*-trimethylammonium chloride (DOTMA). In order to assess the safety of Eph receptor A2 (EphA2) targeting 1,2-dioleoyl-*sn*-glycero-3-phosphocholine (DOPC) encapsulated siRNA (siRNA-EphA2- DOPC) in patients with advanced and recurrent cancer, a phase I clinical trial is now enrolling patients. Cationic polymers like chitosan, cyclodextrin, and polyethyleneimine can be used to concentrate siRNAs (PEI). One of the cyclodextrin polymers conjugated with human transferrin is entering a phase I clinical trial, and its name is CALAA-01. By creating tiny, cationic nanoparticles with the human epidermal growth factor receptor 2 (HER-2 receptor)-specific siRNA, PEI has been employed as an anti-cancer agent. The evaluation of local drug elute R (siG12D LODER), which targets the mutant KRAS oncogene, for the treatment of advanced pancreatic cancer has begun as part of a phase II clinical trial. Enhancing cellular absorption of siRNAs by conjugating to peptides, anti-bodies, and aptamers improves stability during circulation. Nanocarriers have significantly enhanced the targeting specificity, pharmacokinetics, and biodistribution characteristics of siRNAs. Polyallylamine phosphate nanocarriers have been created to disassemble at low endosomal pH and release siRNAs into the cytoplasm.

TABLE 6.2
Summary of Gene Therapy Approaches

S. No.	Gene Therapy	Mechanism of Action	Category	Indication
1	Oncolytic virotherapy (OV)	Directly lyses tumour cells and introduces wild-type tumour suppressor genes into cells	Naturally occurring or genetically modified viruses	Tumour immunotherapy
2	Gendicine	Induces the expression of p53, restores its activity, and destroys the tumour cells	Non-replicative adenoviral vector	Neck and head squamous cell carcinoma
3	Oncorine (rAd5-H101)	Causes oncolysis	Replicative, oncolytic recombinant ad5	Refractory nasopharyngeal cancer
4	Imlygic	Causes apoptosis of tumour cell	Genetically modified oncolytic HSV-1	Non-resectable metastatic melanoma
5	Rexin-G	Inhibits cell cycle in the G1 phase	Replication-incompetent retroviral vector	Metastatic cancers
6	Kymriah	Initiates the anti-tumour effect through CD3 domain	CAR T cell-based gene	Relapsed B-cell acute lymphoblastic leukaemia
7	Zalmoxis	Enhances immune reconstitution	Allogeneic hematopoietic stem cell transplantation (allo-HSCT)	Hematopoietic malignancies

The siRNA-based approach's clinical translation faces difficult problems with dose adjustment, individual variability, and disease stages. Future research will focus on developing the most effective customized therapies and on controlled release to target only particular tumours for treatment. Based on their induction and mode of action, Table 6.2 provides a summary of the gene therapy medications.

6.6 NANOMEDICINE

Due to their small size and high surface-to-volume ratio, nanoparticles (which range in size from 1 to 1,000 nm) have unusual physicochemical characteristics. In cancer treatment, biocompatible nanoparticles are utilized to get around some of the problems with traditional therapy, like the poor specificity and bioavailability of medications or contrast agents. Therefore, the active drugs' solubility and biocompatibility, as well as their stability in body fluids and duration of retention in the tumour vasculature, will all be improved by encapsulating them in nanoparticles. Additionally, by responding to a particular stimulus, nanoparticles can be made to be exceedingly selective for a particular target and release the medicine in a regulated manner. ThermoDox, a liposomal formulation that can release doxorubicin in reaction to a rise in temperature, is an example of this.

For diagnostic applications, inorganic nanoparticles are typically used as contrast agents. Quantum dots are among them; they are tiny, light-emitting semiconductor nanocrystals with unusual electrical and optical characteristics that make them extremely luminous, resistant to photobleaching, and sensitive for detection and imaging. They may be effective theranostic tools when combined with active substances. In a recent work, anti-HER2 antibody was coupled to quantum dots coated with poly(ethylene glycol) (PEG), which were then localized in particular tumour cells.

Due to their interaction with magnetic fields, superparamagnetic iron oxide nanoparticles (SPIONs) are frequently used as contrast agents in magnetic resonance imaging (MRI). Ferumoxides (Feridex in the United States, Endorem in Europe), ferucarbotran (Resovist), ferucarbotran C (Supravist, SHU 555 C), ferumoxtran-10 (Combidex), and NC100150 are five forms of SPIONs that have been studied for MRI (Clariscan). Only a few nations currently sell ferucarbotran; the others have the drug taken off the market. A formulation of iron oxide coated with aminosilane known as Nanotherm has previously been approved for the treatment of glioblastoma, and SPIONs have also been investigated for use in the magnetic hyperthermia method of cancer treatment.

Due to their favourable optical, electrical, and low toxicity characteristics, gold nanoparticles have gained attention. They are primarily employed as contrast agents for computed tomography, photodynamic treatment, photoacoustic imaging, and X-ray imaging. The FDA approved a nanoshell comprised of a silica core and a gold shell coated with PEG in 2012, and it was marketed as AuroShell (Nanospectra) for the photodynamic therapy treatment of breast cancer.

The major purpose of organic nanoparticles is to transport medications. Although both liposomes and micelles are composed of phospholipids, their morphologies are different. Liposomes are spherical, lipid bilayer containing particles that resemble the shape of cell membranes. Hydrophobic medications can either be accommodated in the bilayer or chemically bonded to the particles, although they are primarily utilized to encapsulate hydrophilic pharmaceuticals in their watery core. Instead, micelles have a hydrophobic core that can encapsulate medications that are also hydrophobic. The first nanoparticles to receive FDA approval were called "Doxil," which were PEGylated liposomes loaded with doxorubicin and used to treat Kaposi's sarcoma linked to AIDS. With this formulation, doxorubicin adverse effects are significantly reduced. Other liposomal formulations, including Myocet and DaunoXome, have since received FDA approval for use in the treatment of cancer. Biocompatible or natural polymers, such as poly(lactide-*co*-glycolide), poly(caprolactone), chitosan, alginate, and albumin, are used to make polymeric nanoparticles. The FDA has already approved some formulations, including Ontak (an engineered protein combining interleukin-2 and diphtheria toxins for the treatment of non-peripheral Hodgkin's T-cell lymphomas) and Abraxane (albumin-paclitaxel particles for the treatment of metastatic breast cancer and pancreatic ductal adenocarcinoma).

In addition to these systems, which have either received approval or are currently the subject of clinical evaluation, it is important to mention several fresh nanoparticles that are currently being tested at the research level and should enhance therapy effectiveness. For instance, it has been shown that solid lipid nanoparticles, which are made of lipids that are solid at body temperature and loaded with hydrophobic drugs, provide a higher level of drug stability and prolonged release compared to other systems. However, the encapsulation efficiency is frequently low due to their high crystallinity. One or more liquid at room temperature lipids, such as oleic acid, are added to the formulation to address this problem. Due to their ability to pass the blood–brain barrier, lipid nanoparticles are promising candidates for the treatment of brain tumours. Because they combine the effects of hyperthermia and conventional chemotherapy, lipid nanoparticles loaded with SPIONs and temozolomide are effective for treating glioblastoma, according to a recent study. Another type of nanoparticles known as dendrimers has a globular form and is made of polymers with repeated branching structures. Their structure is incredibly adaptable for a wide range of applications and their design is simple to control. For instance, some recent research demonstrate that doxorubicin-loaded poly-L-lysine (PLL) dendrimers trigger anti-angiogenic responses in *in vivo* cancer models. ImDendrim, a formulation based on a dendrimer and a rhenium complex connected to an imidazolium ligand, is currently the subject of just one clinical trial for the treatment of incurable liver tumours that do not respond to standard therapy.

6.7 NATURAL ANTIOXIDANTS IN CANCER THERAPY

The human body is subjected to a number of exogenous insults on a daily basis, including tobacco smoke, air pollution, and ultraviolet (UV) rays, which cause the body to produce reactive species, particularly oxidants and free radicals, which are in turn responsible for the development of many diseases, including cancer. In addition to being naturally produced inside our cells and tissues by mitochondria and peroxisomes as well as from the metabolism of macrophages during normal physiological aerobic processes, these molecules can also be produced as a result of the clinical administration of medications.

DNA (genetic changes, DNA double-strand breaks, and chromosomal abnormalities) as well as other biomacromolecules, including lipids (membrane peroxidation and necrosis) and proteins, can be harmed by oxidative stress and radical oxygen species (significantly changing the regulation of transcription factors and, as a consequence, of essential metabolic pathways).

Our body contains defence against these chemicals, but they are occasionally unable to stop the significant harm they do. Natural antioxidants like vitamins, polyphenols, and plant-derived bioactive compounds are currently being studied in order to introduce them as preventive agents and potential therapeutic drugs, in addition to research into the roles of the physiological enzymes SOD, CAT, and glutathione peroxidase (GP). These compounds are present in many plants and spices and have anti-inflammatory and antioxidant activities. Vitamins, alkaloids, flavonoids, carotenoids, curcumin, berberine, quercetin, and many more substances have been screened *in vitro* and tested *in vivo* and have been offered as complementary therapy for cancer due to their notable anti-proliferative and pro-apoptotic properties.

Despite the benefits of employing natural medicines, their introduction into clinical settings is still challenging because of their low bioavailability and/or toxicity. Turmeric (*Curcuma longa*) contains a polyphenolic component called curcumin, which has been used as a traditional Southeast Asian medicine for its anti-inflammatory, antioxidant, and therapeutic properties. At the effective therapeutic levels, it has been demonstrated to have cytotoxic effects on a variety of tumour types, including brain, lung, leukaemia, pancreatic, and hepatocellular carcinoma, with no negative effects on normal cells. Numerous cellular pathways can be altered by curcumin, but its biological features, and consequently the length of the course of treatment and the most effective therapeutic doses, are still poorly understood. This molecule is unstable, poorly soluble in water, and highly lipophilic. To increase its bioavailability, various methods and particular carriers, like liposomes and micelles, have been created. There are now 24 active clinical trials using curcumin, and 23 have already been finished.

An alkaloid substance known as berberine is obtained from various plants, including berberis. It has recently been shown to be effective against several tumour types and to act as a chemopreventive drug by altering a number of signalling pathways. It is poorly soluble in water, like curcumin, thus several nanotechnological techniques have been developed to help carry it through cell membranes; six clinical trials are currently open, and one has already been finished.

By attaching to cellular receptors and disrupting numerous signalling pathways, the polyphenolic flavonoid quercetin, which is present in fruits and vegetables, has been shown to be useful in treating a number of tumours, including breast, liver, colon, prostate, and lung cancers. Surprisingly, research has found that it is also effective when used in conjunction with chemotherapy drugs. Seven clinical trials are active at the moment, and four have been finished.

6.8 RECENT INNOVATIONS IN CANCER THERAPY: RADIOMICS AND PATHOMICS

Effective cancer treatment today relies on surgery and, in around 50% of patients, radiotherapy. This treatment can be administered using an external beam source or by inserting a radioactive source locally (this method is known as brachytherapy), which produces focused irradiation. Image-guided radiotherapy (IGRT), in which images of the patient are captured throughout the treatment to allow the best quantity of radiation to be set, now makes it easier to localize the beam. With the advent of intensity-modulated radiotherapy (IMRT), it is now possible to construct radiation fields with varying intensities, assisting in lowering the doses received by healthy tissues and limiting unfavourable side effects. Finally, stereotactic ablative radiotherapy (SABR) has made it possible to provide an ablative dosage of radiation just to a limited target volume, greatly lowering unfavourable effects.

Sadly, radio-resistance can develop throughout treatment, decreasing its effectiveness. This has been connected to mitochondrial abnormalities; therefore, focusing on particular activities has shown to be effective in restoring the anti-cancer benefits. The aberrant shape and size of mitochondria have been associated with radioresistance in a model of oesophageal adenocarcinoma, for instance, and measurements of patients' energy metabolism have made it possible to distinguish between those who are resistant to treatment and those who are susceptible to it. For the treatment of gastrointestinal cancer, small compounds that act as radiosensitizers are being studied that target mitochondria.

Cancer is a complicated disease, and combining the wealth of knowledge discovered through diagnostic and therapeutic processes is essential for its effective treatment. By using non-invasive imaging techniques, an overview of the entire tridimensional volume of the tumour may now be obtained, thanks to the capacity to integrate data from molecular studies and medical imaging. This is in line with precision medicine's primary goal, which is to reduce therapy-related adverse effects while maximizing its effectiveness to produce the greatest individualized therapy.

In order to build up quantitative picture features from radiology and pathology screens as therapeutic and prognostic predictors of illness outcome, two promising and cutting-edge fields have emerged: pathomics and radiomics. In order to handle and elaborate the enormous number of collected data sets and to precisely anticipate the treatment efficacy, the clinical result, and the illness recurrence, many artificial intelligence technologies, such as machine learning applications, have been introduced. Finding an ad hoc adaptation for the best prognosis and outcome can be aided by the prediction of the therapy response. Biomedical pictures are essential to enable real-time monitoring of disease progression since they are strictly associated with cancer molecular characterization. Personalized therapy nowadays necessitates an integrated interpretation of the results produced by numerous diagnostic techniques.

Radiomics is designed to be a high-throughput method of analysing tumour attributes through medical image analysis. On the other hand, pathomics depends on the creation and analysis of high-resolution tissue images. Numerous studies are emphasizing the creation of novel image processing methods in order to extrapolate data through quantification and illness characterization. To detect disease phenotypes, flexible databases are needed to accommodate large amounts of data from gene expression, histology, 3D tissue reconstruction (MRI), and metabolic characteristics (positron emission tomography, PET).

An immediate need exists to specify unambiguous data acquisition guidelines. The German National Cohort Consortium and the Quantitative Imaging Network are two examples of initiatives that have already been launched to create standardized practices and facilitate clinical translation. To establish reliable procedures for the construction and application of computational and statistical approaches, as well as for the acquisition of images, it is necessary to provide a precise description of the parameters. About 50 radiomics clinical trials are presently enrolling patients, and a few are already complete, according to the US National Library of Medicine.

6.9 CONCLUSIONS AND FUTURE PERSPECTIVES

In recent years, research into cancer medicine has taken remarkable steps towards more effective, precise and less invasive cancer. The recent advances in cancer medical research is cited in (Figure 6.1).

While targeted therapy and nanomedicine have improved the biodistribution of new and tested chemotherapy medicines around the targeted tissue, other approaches, including gene therapy, siRNA delivery, immunotherapy, and antioxidant compounds, have given cancer patients additional options. On the other hand, tumour excision methods, including thermal ablation and magnetic hyperthermia, show promise. In order to improve prognosis and outcome, radiomics and pathomics techniques assist in the administration of large data sets from cancer patients.

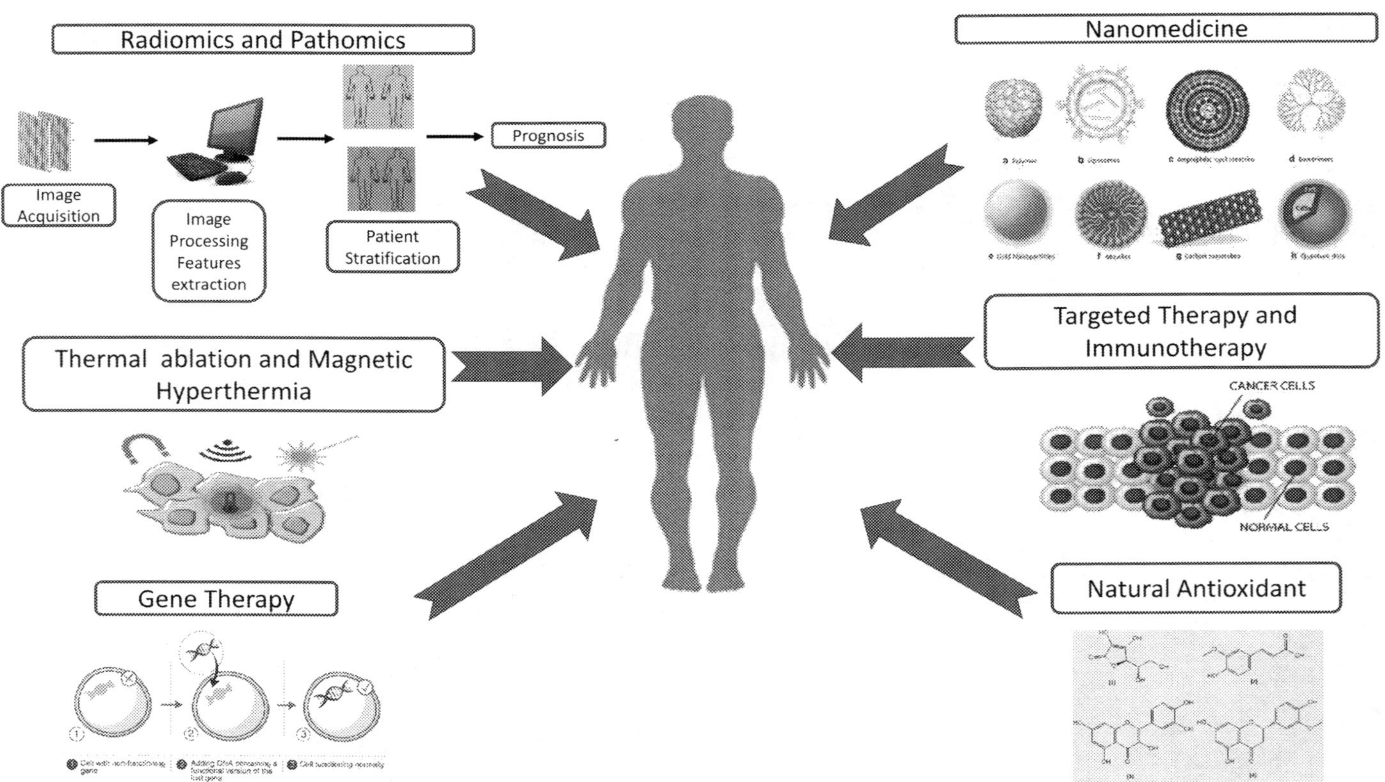

FIGURE 6.1 Recent advancement in cancer medical research.

LIST OF ABBREVIATIONS

ADA	Adenosine deaminase
BM	Bone marrow
CD	Clusters of differentiation
CLL	Chronic lymphocytic leukaemia
CSCs	Cancer stem cells
CXCL	Chemokine ligand
CXCR	Chemokine receptor type
DOPC	1,2-Dioleoyl-*sn*-glycero-3-phosphocholine
DOTAP	*N*-[1-(2,3-dioleoyloxy)propyl]-*N,N,N*-trimethylammonium methyl
DOTMA	*N*-[1-(2, 3-dioleyloxy)propyl]-*N,N,N*-trimethylammonium chloride
EGFR	Epidermal growth factor receptor
EphA2	Ephrin type-A receptor 2
ESCs	Embryonic stem cells
EVs	Extracellular vesicles
FDA	Food and Drug Administration
GP	Glutathione peroxidase
hESCs	Hematopoietic embryonic stem cells
HSCs	Hematopoietic stem cells
IGRT	Image-guided radiotherapy
IL	Interleukin
IMRT	Intensity-modulated radiotherapy
iPSCs	Induced pluripotent stem cells
KRAS	Kirsten rat sarcoma
LFA	Lymphocyte function-associated antigen
LODER	Local drug elute R
MMP	Matrix metalloproteinase
MRI	Magnetic resonance imaging
mRNA	messenger RNA
MSCs	Mesenchymal stem cells
NK	Natural killer
NSCs	Neural stem cells
NSCLC	Non-small cell lung cancer
PEG	Polyethylene glycol
PEI	Polyethylenimine
PET	Positron emission tomography
PKB	Protein kinase B
PLL	Poly-L-lysine
RFA	Radio frequency ablation
RISC	RNA-induced silencing complex
RNAi	RNA interference
SABR	Stereotactic ablative radiotherapy
SCID	Severe combined immune deficiency
SDF	Standing deposit facility
siRNAs	Small interfering RNAs

SPION Superparamagnetic iron oxide nanoparticles
TK Thymidine kinase
TM Tumour microenvironment
VEGF Vascular endothelial growth factor
VLA Very late antigen

BIBLIOGRAPHY

Abadeer, N. S., & Murphy, C. J. (2016). Recent progress in cancer thermal therapy using gold nanoparticles. *The Journal of Physical Chemistry: C*, *120*(9), 4691–4716.

Acharya, S., Dilnawaz, F., & Sahoo, S. K. (2009). Targeted epidermal growth factor receptor nanoparticle bioconjugates for breast cancer therapy. *Biomaterials*, *30*(29), 5737–5750.

Aerts, H. J. W. L. (2016). The potential of radiomic-based phenotyping in precision medicine a review. *JAMA Oncology*, *2*(12), 1636–1642.

Aerts, H. J. W. L., Velazquez, E. R., Leijenaar, R. T. H., et al. (2014) Decoding tumour phenotype by noninvasive imaging using a quantitative radiomics approach. *Nature Communications*, *5*, 4006

Ahrendt, S. A., Hu, Y., Buta, M., et al. (2003). p53 mutations and survival in stage I non-small-cell lung cancer: Results of a prospective study. *Journal of the National Cancer Institute*, *95*(13), 961–970.

Alasvand, N., Urbanska, A. M., Rahmati, M., et al. (2017) Therapeutic nanoparticles for targeted delivery of anticancer drugs. *Multifunctional systems for combined delivery, biosensing and diagnostics* (1st ed., Chapter 13, pp. 245–259), Elsevier.

Albanese, A., Tang, P. S., & Chan, W. C. W. (2012). The effect of nanoparticle size, shape, and surface chemistry on biological systems. *Annual Review of Biomedical Engineering*, *14*, 1–16.

Al-Jamal, K. T., Al-Jamal, W. T., Wang, T. W., et al. (2013). Cationic poly-L-lysine dendrimer complexes doxorubicin and delays tumor growth *in vitro* and *in vivo*. *ACS Nano*, *7*(3), 1905–1917.

Allenson, K., Castillo, J., Lucas, S., et al. (2017). High prevalence of mutant KRAS in circulating exosome-derived DNA from early-stage pancreatic cancer patients. *Annals of Oncology*, *28*(4), 741–747.

Bagalkot, V., Zhang, L., Levy-Nissenbaum, E., et al. (2007). Quantum dot-aptamer conjugates for synchronous cancer imaging, therapy, and sensing of drug delivery based on bifluorescence resonance energy transfer. *Nano Letters*, *7*(10), 3065–3070.

Barenholz, Y. (2012). Doxil—the first FDA-approved nano-drug: Lessons learned. *Journal of Controlled Release*, *160*(2), 117–134.

Barua, S., & Mitragotri, S. (2014). Challenges associated with penetration of nanoparticles across cell and tissue barriers: A review of current status and future prospects. *Nano Today*, *9*(2), 223–243.

Bazak, R., Houri, M., Achy, E. S., et al. (2015). Cancer active targeting by nanoparticles: A comprehensive review of literature. *Journal of Cancer Research and Clinical Oncology*, *141*(5), 769–784.

Bernardini, S., Tiezzi, A., Masci, L., et al. (2018). Natural products for human health: An historical overview of the drug discovery approaches. *Natural Product Research*, *32*(16), 1926–1950.

Besse, B., Charrier, M., Lapierre, V., et al. (2016). Dendritic cell-derived exosomes as maintenance immunotherapy after first line chemotherapy in NSCLC. *Oncoimmunology*, *5*(4), e1071008.

Bianchi, S., & Giovannini, L. (2018). Inhibition of mTOR/S6K1/4E-BP1 signaling by nutraceutical sirt1 modulators. *Nutrition and Cancer*, *70*(3), 490–501.

Bora, R. S., Gupta, D., Mukkur, T. K. S., et al. (2012). RNA interference therapeutics for cancer: Challenges and opportunities (review). *Molecular Medicine Reports*, *6*(1), 9–15.

Boriachek, K., Islam, M. N., Möller, A., et al. (2018). Biological functions and current advances in isolation and detection strategies for exosome nanovesicles. *Small*, *14*(6), 1702153.

Boyer, C., Whittaker, M. R., Bulmus, V., et al. (2010) The design and utility of polymer-stabilized iron-oxide nanoparticles for nanomedicine applications.NPG *Asia Materials*, 2, 23–30.

Brace, C. (2011). Thermal tumor ablation in clinical use. *IEEE Pulse*, *2*(5), 28–38.

Bregoli, L., Movia, D., Gavigan-Imedio, J. D., et al. (2016). Nanomedicine applied to translational oncology: A future perspective on cancer treatment. *Nanomedicine*, *12*(1), 81–103.

Brito, A. F., Ribeiro, M., Abrantes, A. M., et al. (2015). Quercetin in cancer treatment, alone or in combination with conventional therapeutics? *Current Medicinal Chemistry*, *22*(26), 305–339.

Byrne, J. D., Betancourt, T., Brannon-Peppas, L., et al. (2008). Active targeting schemes for nanoparticle systems in cancer therapeutics. *Advanced Drug Delivery Reviews*, *60*(15), 1615–1626.

Cadet, J., Douki, T., & Ravanat, J. L. (1997). Artifacts associated with the measurement of oxidized DNA bases. *Environment Health Perspectives*, *105*(10), 1034–1039.

Cazzoli, R., Buttitta, F., Di Nicola, M., et al (2013). MicroRNAs derived from circulating exosomes as noninvasive biomarkers for screening and diagnosing lung cancer. *Journal of Thoracic Oncology*, *8*(9), 1156–1162.

Chikara, S., Nagaprashantha, L. D., & Singhal, J. (2018). Oxidative stress and dietary phytochemicals: Role in cancer chemoprevention and treatment. *Cancer Letters*, *413*, 122–134.

Cho, J. A., Yeo, D. J., Son, H. Y., et al. (2005). Exosomes: A new delivery system for tumor antigens in cancer immunotherapy. *International Journal of Cancer*, *114*(4), 613–622.

Colombo, M., Raposo, G., & Théry, C. (2014). Biogenesis, secretion, and intercellular interactions of exosomes and other extracellular vesicles. *Annual Review of Cell and Developmental Biology*, *30*, 255–289.

Connor, E. E., Mwamuka, J., Gole, A., et al. (2005). Gold nanoparticles are taken up by human cells but do not cause acute cytotoxicity. *Small*, *1*(3), 325–327.

Cornelio, D. B., Roesler, R., & Schwartsmann, G. (2007). Gastrin-releasing peptide receptor as a molecular target in experimental anticancer therapy. *Annals of Oncology*, *18*(9), 1457–1466.

Costa-Silva, B., Aiello, N. M., Ocean, A. J., et al. (2015). Pancreatic cancer exosomes initiate pre-metastatic niche formation in the liver. *Nature Cell Biology*, *17*(6), 816–826.

Czauderna, F., Fechtner, M., Dames, S., et al. (2003). Structural variations and stabilising modifications of synthetic siRNAs in mammalian cells. *Nucleic Acids Research*, *31*(11), 2705–2716.

Dagogo-Jack, I., & Shaw, A. T. (2018). Tumour heterogeneity and resistance to cancer therapies. *Nature Reviews Clinical Oncology*, *15*(2), 81–94.

Dai, S., Wei, D., Wu, Z., et al. (2008). Phase I clinical trial of autologous ascites-derived exosomes combined with GM-CSF for colorectal cancer. *Molecular Therapy*, *16*(4), 782–790.

Daniels, T. R., Bernabeu, E., Rodríguez, J. A., et al. (2012). The transferrin receptor and the targeted delivery of therapeutic agents against cancer. *Biochimica et Biophysica Acta*, *1820*(3), 291–317.

Del Re, M., Biasco, E., Crucitta, S., et al. (2017). The detection of androgen receptor splice variant 7 in plasma-derived exosomal RNA strongly predicts resistance to hormonal therapy in metastatic prostate cancer patients. *European Urology*, *71*(4), 680–687.

Demeule, M., Currie, J. C., Bertrand, Y., et al. (2008). Involvement of the low-density lipoprotein receptor-related protein in the transcytosis of the brain delivery vector angiopep-2. *Journal of Neurochemistry*, *106*(4), 1534–1544.

Elbashir, S. M., Harborth, J., Lendeckel, W., et al. (2001). Duplexes of 21-nucleotide RNAs mediate RNA interference in cultured mammalian cells. *Nature*, *411*(6836), 494–498.

Elouahabi, A., & Ruysschaert, J. M. (2005). Formation and intracellular trafficking of lipoplexes and polyplexes. *Molecular Therapy*, *11*(3), 336–347.

Farokhzad, O. C., Cheng, J., Teply, B. A., et al. (2006). Targeted nanoparticle–aptamer bioconjugates for cancer chemotherapy *in vivo*. *Proceedings of the National Academy of Sciences of the United States of America*, *103*(16), 6315–20.

Farooqi, A. A., Qureshi, M. Z., Khalid, S., et al. (2019). Regulation of cell signaling pathways by berberine in different cancers: Searching for missing pieces of an incomplete jig-saw puzzle for an effective cancer therapy. *Cancers (Basel)*, *11*(4), E478.

Fire, A. Z. (2007). Gene silencing by double-stranded RNA. *Cell Death and Differentiation*, *14*(12), 1998–2012.

Fleming, J. B. (2005). Molecular consequences of silencing mutant K-ras in pancreatic cancer cells: Justification for K-ras-directed therapy. *Molecular Cancer Research*, *3*(7), 413–423.

Floyd, R. A., Watson, J. J., Wong, P. K., et al. (1986). Hydroxyl free radical adduct of deoxyguanosine: Sensitive detection and mechanisms of formation. *Free Radical Research Communications*, *1*(3), 163–172.

Freeman, S. M., Abboud, C., Freeman, S. M., et al. (1993). The "bystander effect": Tumor regression when a fraction of the tumor mass is genetically modified. *Cancer Research*, *53*(21), 5274–5283.

Friedmann, T. (1992). A brief history of gene therapy. *Nature Genetics*, *2*(2), 93–98.

Ganesan, P., & Narayanasamy, D. (2017). Lipid nanoparticles: Different preparation techniques, characterization, hurdles, and strategies for the production of solid lipid nanoparticles and nanostructured lipid carriers for oral drug delivery. *Sustainable Chemistry and Pharmacy*, *6*, 37–56.

Gao, J., Chen, K., Miao, Z., et al. (2011). Affibody-based nanoprobes for HER2-expressing cell and tumor imaging. *Biomaterials*, *32*(8), 2141–2148.

Gerlowski, L. E., & Jain, R. K. (1986). Microvascular permeability of normal and neoplastic tissues. *Microvascular Research*, *31*(3), 288–305.

Gille, G., & Sigler, K. (1995). Oxidative stress and living cells. *Folia Microbiolica (Praha)*, *40*(2), 131–152.

Gillies, E. R., & Fréchet, J. M. J. (2005). Dendrimers and dendritic polymers in drug delivery. *Drug Discovery Today*, *10*(1), 35–43.

Ginn, S. L., Amaya, A. K., Alexander, I. E., et al. (2018). Gene Therapy clinical trials worldwide to 2017: An update. *Journal of Gene Medicine*, *20*(5), e3015.

González-Vallinas, M., González-Castejón, M., Rodríguez-Casado, A., et al. (2013). Dietary phytochemicals in cancer prevention and therapy: A complementary approach with promising perspectives. *Nutrition Reviews*, *71*(9), 585–599.

Gray, M. J., Van Buren, G., Dallas, N. A., et al. (2008). Therapeutic targeting of neuropilin-2 on colorectal carcinoma cells implanted in the murine liver. *Journal of the National Cancer Institute*, *100*(2), 109–120.

Griffith, T. S. T., Stokes, B., Kucaba, T. A., et al. (2009). TRAIL gene therapy: From preclinical development to clinical application. *Current Gene Therapy*, *9*(1), 9–19.

Grillone, A., Riva, E. R., Mondini, A., et al. (2015). Active targeting of sorafenib: Preparation, characterization, and *in vitro* testing of drug-loaded magnetic solid lipid nanoparticles. *Advanced Healthcare Materials*, *4*(11), 1681–1690.

Grove, O., Berglund, A. E., Schabath, M. B., et al. (2015). Quantitative computed tomographic descriptors associate tumor shape complexity and intratumor heterogeneity with prognosis in lung adenocarcinoma. *PLoS One*, *10*(3), e0118261.

Gubernator, J. (2011). Active methods of drug loading into liposomes: Recent strategies for stable drug entrapment and increased *in vivo* activity. *Expert Opinion on Drug Delivery*, *8*(5), 565–580.

Gupta, R. K., Patel, A. K., Shah, N., et al. (2014). Oxidative stress and cancer: An overview. *Asian Pacific Journal of Cancer Prevention*, *15*(11), 4405–4409.

Hadla, M., Palazzolo, S., & Corona, G. (2016). Exosomes increase the therapeutic index of doxorubicin in breast and ovarian cancer mouse models. *Nanomedicine (London)*, *11*(18), 2431–2441.

Halder, J., Kamat, A. A., Landen, C. N., et al. (2006). Focal adhesion kinase targeting using *in vivo* short interfering RNA delivery in neutral liposomes for ovarian carcinoma therapy. *Clinical Cancer Research*, *12*(16), 4916–4924.

Hall, A. H. S., Wan, J., Shaughnessy, E. E., et al. (2004). RNA interference using boranophosphate siRNAs: Structure–activity relationships. *Nucleic Acids Research*, *32*(20), 5991–6000.

Halliwell, B. (2006). Oxidative stress and cancer: Have we moved forward? *The Biochemical Journal*, *401*(1), 1–11.

Hervault, A., & Thanh, N. T. K. (2014). Magnetic nanoparticle-based therapeutic agents for thermo-chemotherapy treatment of cancer. *Nanoscale*, *6*(20), 11553–11573.

Hofheinz, R. D., Gnad-Vogt, S. U., Beyer, U., et al. (2005). Liposomal encapsulated anticancer drugs. *Anticancer Drugs*, *16*(7), 691–707.

Hornung, V., Guenthner-Biller, M., Bourquin, C., et al. (2005). Sequence-specific potent induction of IFN-α by short interfering RNA in plasmacytoid dendritic cells through TLR7. *Nature Medicine*, *11*(3), 263–270.

Huang, S., Li, J., Han, L., et al. (2011). Dual targeting effect of angiopep-2-modified, DNA-loaded nanoparticles for glioma. *Biomaterials*, *32*(28), 6832–6838.

Hymel, D., & Peterson, B. R. (2012). Synthetic cell surface receptors for delivery of therapeutics and probes. *Advanced Drug Delivery Reviews*, *64*(9), 797–810.

Imran, M., Ullah, A., Saeed, F., et al. (2018). Cucurmin, anticancer, & antitumor perspectives: A comprehensive review. *Critical Reviews in Food Science and Nutrition*, *58*(8), 1271–1293.

Iqbal, J., Abbasi, B. A., Mahmood, T., et al. (2017). Plant-derived anticancer agents: A green anticancer approach. *Asian Pacific Journal of Tropical Biomedicine*, *7*(12), 1129–115.

Jackson, A. L., & Linsley, P. S. (2010). Recognizing and avoiding siRNA off-target effects for target identification and therapeutic application. *Nature Reviews Drug Discovery*, *9*(1), 57–67.

Jang, S. C., Kim, O. Y., Yoon, C. M., et al. (2013). Bioinspired exosome-mimetic nanovesicles for targeted delivery of chemotherapeutics to malignant tumors. *ACS Nano*, *7*(9), 7698–7710.

Jia, L. T., Chen, S. Y., & Yang, A. G. (2012). Cancer gene therapy targeting cellular apoptosis machinery. *Cancer Treatment Reviews*, *38*(7), 868–876.

Judge, A. D., Sood, V., Shaw, J. R., et al. (2005). Sequence-dependent stimulation of the mammalian innate immune response by synthetic siRNA. *Nature Biotechnology*, *23*(4), 457–462.

Kabary, D. M., Helmy, M. W., Abdelfattah, E.-Z. A., et al. (2018) Inhalable multi-compartmental phospholipid enveloped lipid core nano-composites for localized mTOR inhibitor/herbal combined therapy of lung carcinoma. *European Journal of Pharmaceutics and Biopharmaceutics*, *130*, 152–164.

Kahlert, C., Melo, S. A., Protopopov, A., et al. (2014). Identification of double stranded genomic DNA spanning all chromosomes with mutated KRAS and P53 DNA in the serum exosomes of patients with pancreatic cancer. *Journal of Biological Chemistry, 289*(7), 3869–3875.

Katz, L., & Baltz, R. H. (2016). Natural product discovery: Past, present, and future. *Journal of Industrial Microbiology and Biotechnology, 43*(2–3), 155–176.

Kenny, G. D., Kamaly, N., Kalber, T. L., et al. (2011). Novel multifunctional nanoparticle mediates siRNA tumour delivery, visualisation and therapeutic tumour reduction *in vivo*. *Journal of Controlled Release, 149*(2), 111–116.

Kesharwani, P., Jain, K., & Jain, N. K. (2014). Dendrimer as nanocarrier for drug delivery. *Progress in Polymer Science, 39*(2), 268–307.

Kim, E. M., & Jeong, H. J. (2017). Current status and future direction of nanomedicine: Focus on advanced biological and medical applications. *Nuclear Medicine and Molecular Imaging, 51*(2), 106–117.

Kim, M. S., Haney, M. J., Zhao, Y., et al. (2016). Development of exosome-encapsulated paclitaxel to overcome MDR in cancer cells. *Nanomedicine, 12*(3), 655–664.

Kim, M. S., Haney, M. J., Zhao, Y., et al. (2018). Engineering macrophage-derived exosomes for targeted paclitaxel delivery to pulmonary metastases: *In vitro* and *in vivo* evaluations. *Nanomedicine, 14*(1), 195–204.

Kim, H. S., Song, I. H., Kim, J. C., et al. (2006). *In vitro* and *in vivo* gene-transferring characteristics of novel cationic lipids, DMKD (*o,o′*- dimyristyl-*n*-lysyl aspartate) and DMKE (*o,o′*-dimyristyl-*n*-lysyl glutamate). *Journal of Controlled Release, 115*(2), 234–241.

Kocaadam, B., & Şanlier, N. (2017). Curcumin, an active component of turmeric (*Curcuma longa*), and its effects on health. *Critical Reviews in Food Science and Nutrition, 57*(13), 2889–2895.

Kong, J., Cooper, L. A. D., Wang, F., et al. (2013). Machine-based morphologic analysis of glioblastoma using whole-slide pathology images uncovers clinically relevant molecular correlates. *PLoS One, 8*(11), e81049.

Kosaka, N., Urabe, F., Egawa, S., et al. (2017). The small vesicular culprits: The investigation of extracellular vesicles as new targets for cancer treatment. *Clinical and Translational Medicine, 6*(1), 45.

Kreuter, J., Ramge, P., Petrov, V., et al. (2003). Direct evidence that polysorbate-80-coated poly(butylcyanoacrylate) nanoparticles deliver drugs to the CNS via specific mechanisms requiring prior binding of drug to the nanoparticles. *Pharmaceutical Research, 20*(3), 409–416.

Kulhari, H., Pooja, D., & Shrivastava, S. (2014). Peptide conjugated polymeric nanoparticles as a carrier for targeted delivery of docetaxel. *Colloids and Surfaces B: Biointerfaces, 117*, 166–173.

Kumar, B., Garcia, M., Murakami, J. L., et al. (2016). Exosome-mediated microenvironment dysregulation in leukemia. *Biochimica et Biophysica Acta, 1863*(3), 464–470.

Kumar, G., Mittal, S., & Sak, K. (2016). Molecular mechanisms underlying chemopreventive potential of curcumin: Current challenges and future perspectives. *Life Sciences, 148*, 312–328.

Kunnumakkara, A. B., Bordoloi, D., Harsha, C., et al. (2017). Curcumin mediates anticancer effects by modulating multiple cell signaling pathways. *Clinical Science (London), 131*(15), 1781–1799.

Lapierre, V., Théry, C., Virault-Rocroy, P., et al. (2010). Updated technology to produce highly immunogenic dendritic cell-derived exosomes of clinical grade. *Journal of Immunotherapy, 34*(1), 65–75.

Lebedeva, I. V., Su, Z. Z., Sarkar, D., et al. (2003). Restoring apoptosis as a strategy for cancer gene therapy: Focus on p53 and mda-7. *Seminars in Cancer Biology, 13*(2), 169–178.

Leiner, T., Gerretsen, S., Botnar, R., et al. (2005). Magnetic resonance imaging of atherosclerosis. *European Radiology, 15*(6), 1087–1099.

Liao, H., & Wang, J. H. (2005). Biomembrane-permeable and ribonuclease-resistant siRNA with enhanced activity. *Oligonucleotides, 15*(3), 196–205.

Liu, Y., Tang, Z. G., Lin, Y., et al. (2017) Effects of quercetin on proliferation and migration of human glioblastoma U251 cells. *Biomedicine & Pharmacotherapy*, 92, 33–38.

Liu, J., Xiao, Y., & Allen, C. (2004). Polymer–drug compatibility: A guide to the development of delivery systems for the anticancer agent, ellipticine. *Journal of Pharmaceutical Sciences, 93*(1), 132–143.

Liu, W., Zhai, Y., Heng, X., et al. (2016). Oral bioavailability of curcumin: Problems and advancements. *Journal of Drug Target, 24*(8), 694–702.

Liu, T., Zhang, X., Gao, S., et al. (2016). Exosomal long noncoding RNA CRNDE-h as a novel serum-based biomarker for diagnosis and prognosis of colorectal cancer. *Oncotarget, 7*(51), 85551–85563.

Luga, V., & Wrana, J. L. (2013). Tumor–stroma interaction: Revealing fibroblast-secreted exosomes as potent regulators of Wnt-planar cell polarity signaling in cancer metastasis. *Cancer Research, 73*(23), 6843–6847.

Maeda, H. (2015). Toward a full understanding of the EPR effect in primary and metastatic tumors as well as issues related to its heterogeneity. *Advanced Drug Delivery Reviews, 91*, 3–6.

Malam, Y., Loizidou, M., & Seifalian, A. M. (2009). Liposomes and nanoparticles: Nanosized vehicles for drug delivery in cancer. *Trends in Pharmacological Sciences, 30*(11), 592–599.

Markman, M. (2006). Pegylated liposomal doxorubicin in the treatment of cancers of the breast and ovary. *Expert Opinion on Pharmacotherapy, 7*(11), 1469–1474.

Martina, M. S., Nicolas, V., Wilhelm, C., et al. (2007). The *in vitro* kinetics of the interactions between PEG-ylated magnetic-fluid-loaded liposomes and macrophages. *Biomaterials, 28*(28), 4143–4153.

Martinelli, C. (2017) Exosomes: New biomarkers for targeted cancer therapy. *Molecular oncology: Underlying mechanisms and translational advancements* (1st ed., chapter 6, pp. 129–157). Springer Nature.

Martinelli, C., Pucci, C., & Ciofani, G. (2019). Nanostructured carriers as innovative tools for cancer diagnosis and therapy. *APL Bioengineering, 3*(1), 011502.

Matea, C. T., Mocan, T., Tabaran, F., et al. (2017) Quantum dots in imaging, drug delivery and sensor applications. *International Journal of Nanomedicine, 12*, 5421–5431.

May, J. P., & Li, S.-D. (2013). Hyperthermia-induced drug targeting. *Expert Opinion on Drug Delivery, 10*(4), 511–527.

McCune, J. S. (2018). Rapid advances in immunotherapy to treat cancer. *Clinical Pharmacology & Therapeutics, 103*(4), 540–544.

McKiernan, J., Donovan, M. J., O'Neill, V., et al. (2016). A novel urine exosome gene expression assay to predict high-grade prostate cancer at initial biopsy. *JAMA Oncology, 2*(7), 882–889.

Melo, S. A., Luecke, L. B., Kahlert, C., et al. (2015). Glypican-1 identifies cancer exosomes and detects early pancreatic cancer. *Nature, 523*(7559), 177–182.

Miliotou, A. N., & Papadopoulou, L. C. (2018). CAR t-cell therapy: A new era in cancer immunotherapy. *Current Pharmaceutical Biotechnology, 19*(1), 5–18.

Mizrak, A., Bolukbasi, M. F., Ozdener, G. B., et al. (2013). Genetically engineered microvesicles carrying suicide mRNA/protein inhibit schwannoma tumor growth. *Molecular Therapy, 21*(1), 101–108.

Mohammadinejad, R., Ahmadi, Z., Tavakol, S., et al. (2019) Berberine as a potential autophagy modulator. *Journal of Cell Physiology, 234*(9), 14914–14926.

Muhamad, N., Plengsuriyakarn, T., & Na-Bangchang, K. (2018). Application of active targeting nanoparticle delivery system for chemotherapeutic drugs and traditional/herbal medicines in cancer therapy: A systematic review. *International Journal of Nanomedicine*, *13*, 3921–3935.

Müller, R. H., Radtke, M., & Wissing, S. A. (2002). Nanostructured lipid matrices for improved microencapsulation of drugs. *International Journal of Pharmaceutics*, *242*(1–2), 121–128.

Murakami, A., Ashida, H., & Terao, J. (2008). Multitargeted cancer prevention by quercetin. *Cancer Letters*, *269*(2), 315–325.

Narang, A. S., Delmarre, D., & Gao, D. (2007). Stable drug encapsulation in micelles and microemulsions. *International Journal of Pharmaceutics*, *345*(1–2), 9–25.

Nasir, A., Kausar, A., & Younus, A. (2015). A review on preparation, properties and applications of polymeric nanoparticle-based materials. *Polymer–Plastics Technology and Engineering*, *54*(4), 325–341.

Nasu, Y., Saika, T., Ebara, S., et al. (2007). Suicide gene therapy with adenoviral delivery of HSV-tK gene for patients with local recurrence of prostate cancer after hormonal therapy. *Molecular Therapy*, *15*(4), 834–840.

Natsume, A., & Yoshida, J. (2008). Gene therapy for high-grade glioma: Current approaches and future directions. *Cell Adhesion & Migration*, *2*(3), 186–191.

Ni, X., Castanares, M., Mukherjee, A., et al. (2011). Nucleic acid aptamers: Clinical applications and promising new horizons. *Current Medicinal Chemistry*, *18*(27), 4206–4214.

Ohno, S. I., Takanashi, M., Sudo, K., et al. (2013). Systemically injected exosomes targeted to EGFR deliver antitumor microRNA to breast cancer cells. *Molecular Therapy*, *21*(1), 185–191.

Perrone, D., Ardito, F., Giannatempo, G., et al. (2015). Biological and therapeutic activities, and anticancer properties of curcumin. *Experimental and Therapeutic Medicine*, *10*(5), 1615–1623.

Putney, S. D., Brown, J., Cucco, C., et al. (1999). Enhanced anti-tumor effects with microencapsulated c-myc antisense oligonucleotide. *Antisense & Nucleic Acid Drug Development*, *9*(5), 451–458.

Qi, H., Liu, C., Long, L., et al. (2016). Blood exosomes endowed with magnetic and targeting properties for cancer therapy. *ACS Nano*, *10*(3), 3323–3333.

Rahimi, H. R., Nedaeinia, R., Sepehri Shamloo, A., et al. (2016). Novel delivery system for natural products: Nano-curcumin formulations. *Avicenna Journal of Phytomedicine*, *6*(4), 383–398.

Raimondo, S., Saieva, L., Corrado, C., et al. (2015) Chronic myeloid leukemia-derived exosomes promote tumor growth through an autocrine mechanism *Cell Communication and Signaling*, *13*,8.

Raty, J., Pikkarainen, J., Wirth, T., et al. (2010). Gene therapy: The first approved gene-based medicines, molecular mechanisms and clinical indications. *Current Molecular Pharmacology*, *1*(1), 13–23.

Recht, L., Torres, C. O., Smith, T. W., et al. (1990). Transferrin receptor in normal and neoplastic brain tissue: Implications for brain-tumor immunotherapy. *Journal of Neurosurgery*, *72*(6), 941.

Rosenberg, S. A., Aebersold, P., Cornetta, K., et al. (1990). Gene transfer into humans: Immunotherapy of patients with advanced melanoma, using tumor-infiltrating lymphocytes modified by retroviral gene transduction. *The New England Journal of Medicine*, *323*(9), 570–578.

Rosenberg, S. A., Restifo, N. P., Yang, J. C., et al. (2008). Adoptive cell transfer: A clinical path to effective cancer immunotherapy. *Nature Reviews Cancer*, *8*(4), 299–308.

Rossi, J. J. (2006). RNAi therapeutics: SNALPing siRNAs *in vivo*. *Gene Therapy*, *13*(7), 583–584.

Roth, J. A., Nguyen, D., Lawrence, D. D., et al. (1996). Retrovirus-mediated wild-type p53 gene transfer to tumors of patients with lung cancer. *Nature Medicine*, *2*(9), 985–991.

Sanchez, C., Belleville, P., Popall, M., et al. (2011). Applications of advanced hybrid organic–inorganic nanomaterials: From laboratory to market. *Chemical Society Reviews*, *40*(2), 696–753.

Sarisozen, C., Salzano, G., & Torchilin, V. P. (2015). Recent advances in siRNA delivery. *Biomolecular Concepts*, *6*(5–6), 321–341.

Scanlon, K. (2005). Anti-genes: Sirna, ribozymes and antisense. *Current Pharmaceutical Biotechnology*, *5*(5), 415–420.

Senol, S., Ceyran, A. B., Aydin, A., et al. (2015). Folate receptor α expression and significance in endometrioid endometrium carcinoma and endometrial hyperplasia. *International Journal of Clinical and Experimental Pathology*, *8*(5), 5633–5641.

Shanker, M., Jin, J., Branch, C. D., et al. (2011) Tumor suppressor gene-based nanotherapy: From test tube to the clinic. *Journal of Drug Delivery*, *2011*, 465845.

Sharkey, R. M., & Goldenberg, D. M. (2009). Targeted therapy of cancer: New prospects for antibodies and immunoconjugates. *CA: A Cancer Journal for Clinicians*, *56*(4), 226–243.

Shen, R., Kim, J. J., Yao, M., et al. (2016) Development and evaluation of vitamin E D-α-tocopheryl polyethylene glycol 1000 succinate-mixed polymeric phospholipid micelles of berberine as an anticancer nanopharmaceutical. *International Journal of Nanomedicine*, 11, 1687–1700.

Shi, J., Kantoff, P. W., Wooster, R., et al. (2017). Cancer nanomedicine: Progress, challenges and opportunities. *Nature Reviews Cancer*, *17*(1), 20–37.

Shi, J., Votruba, A. R., Farokhzad, O. C., et al. (2010). Nanotechnology in drug delivery and tissue engineering: From discovery to applications. *Nano Letters*, *10*(9), 3223–3230.

Shih, H., Pickwell, G. V., & Quattrochi, L. C. (2000). Differential effects of flavonoid compounds on tumor promoter-induced activation of the human CYP1A2 enhancer. *Archives of Biochemistry and Biophysics*, *373*(1), 287–294.

Simpson, R. J., Lim, J. W. E., Moritz, R. L., et al. (2009). Exosomes: Proteomic insights and diagnostic potential. *Expert Review of Proteomics*, *6*(3), 267–283.

Singh, S., Sharma, B., Kanwar, S. S., et al. (2016) Lead phytochemicals for anticancer drug development. *Frontiers in Plant Science*, *7*, 1667.

Sinha, R. (2006). Nanotechnology in cancer therapeutics: Bioconjugated nanoparticles for drug delivery. *Molecular Cancer Therapeutics*, *5*(8), 1909–1917.

Siravegna, G., Marsoni, S., Siena, S., et al. (2017). Integrating liquid biopsies into the management of cancer. *Nature Reviews Clinical Oncology*, *14*(9), 531–548.

Skotland, T., Ekroos, K., Kauhanen, D., et al. (2017) Molecular lipid species in urinary exosomes as potential prostate cancer biomarkers. *European Journal of Cancer*, *70*, 122–132.

Sordillo, P. P., & Helson, L. (2015). Curcumin and cancer stem cells: Curcumin has asymmetrical effects on cancer and normal stem cells. *Anticancer Research*, *35*(2), 599–614.

Soutschek, J., Akinc, A., Bramlage, B., et al. (2004). Therapeutic silencing of an endogenous gene by systemic administration of modified siRNAs. *Nature*, *432*(7014), 173–178.

Stuchinskaya, T., Moreno, M., Cook, M. J., et al. (2011). Targeted photodynamic therapy of breast cancer cells using antibody-phthalocyanine-gold nanoparticle conjugates. *Photochemical & Photobiological Sciences*, *10*(5), 822–831.

Suetsugu, A., Honma, K., Saji, S., et al. (2013). Imaging exosome transfer from breast cancer cells to stroma at metastatic sites in ortho-topic nude-mouse models. *Advanced Drug Delivery Reviews*, *65*(3), 383–390.

Sun, T., Zhang, Y. S., Pang, B., et al. (2014). Engineered nanoparticles for drug delivery in cancer therapy. *Angewandte Chemie, International Edition*, *53*(46), 12320–12464.

Sun, D., Zhuang, X., Xiang, X., et al. (2010). A novel nanoparticle drug delivery system: The anti-inflammatory activity of curcumin is enhanced when encapsulated in exosomes. *Molecular Therapy*, *18*(9), 1606–1614.

Tao, W., Zhang, J., Zeng, X., et al. (2015). Blended nanoparticle system based on miscible structurally similar polymers: A safe, simple, targeted, and surprisingly high efficiency vehicle for cancer therapy. *Advanced Healthcare Materials*, *4*(8), 1203–1214.

Tapeinos, C., Marino, A., Battaglini, M., et al. (2018). Stimuli-responsive lipid-based magnetic nanovectors increase apoptosis in glioblastoma cells through synergic intracellular hyperthermia and chemotherapy. *Nanoscale*, *11*(1), 72–88.

Thakur, B. K., Zhang, H., Becker, A., et al. (2014). Double-stranded DNA in exosomes: A novel biomarker in cancer detection. *Cell Research*, *24*(6), 766–769.

Thery, C. (2011). Exosomes: Secreted vesicles and intercellular communications. *F1000 Biology Reports*, *3*, 15.

Tinkle, S., Mcneil, S. E., Mühlebach, S., et al. (2014) Nanomedicines: Addressing the scientific and regulatory gap. *Annals of the New York Academy of Sciences*, *1313*, 35–56.

Ulbrich, K., Hekmatara, T., Herbert, E., et al. (2009). Transferrin- and transferrin-receptor-antibody-modified nanoparticles enable drug delivery across the blood–brain barrier (BBB). *European Journal of Pharmaceutics and Biopharmaceutics*, *71*(2), 251–256.

Vaishnaw, A. K., Gollob, J., Gamba-Vitalo, C., et al. (2010). A status report on RNAi therapeutics. *Silence*, *1*(1), 14.

Valadi, H., Ekström, K., Bossios, A., et al. (2007). Exosome-mediated transfer of mRNAs and microRNAs is a novel mechanism of genetic exchange between cells. *Nature Cell Biology*, *9*(6), 654–659.

Van De Water, F. M., Boerman, O. C., Wouterse, A. C., et al. (2006). Intravenously administered short interfering RNA accumulates in the kidney and selectively suppresses gene function in renal proximal tubules. *Drug Metabolism and Disposition*, *34*(8), 1393–1397.

Vita, M., & Henriksson, M. (2006). The myc oncoprotein as a therapeutic target for human cancer. *Seminars in Cancer Biology*, *16*(4), 318–330.

Vlassov, A. V., Magdaleno, S., Setterquist, R., et al. (2012). Exosomes: Current knowledge of their composition, biological functions, and diagnostic and therapeutic potentials. *Biochimica et Biophysica Acta*, *1820*(7), 940–948.

Waghmare, A. S., Grampurohit, N. D., Gadhave, M. V., et al. (2012). Solid lipid nanoparticles: A promising drug delivery system. *International Research Journal of Pharmacy*, *3*(4), 100–107.

Wang, Y.-J., Pan, M.-H., Cheng, A. L., et al. (1997). Stability of curcumin in buffer solutions and characterization of its degradation products. *Journal of Pharmaceutical and Biomedical Analysis*, *15*(12), 1867–1876.

Wang, Z. P., Wu, J. B., Chen, T.-S., et al. (2015) *In vitro* and *in vivo* antitumor efficacy of berberine-nanostructured lipid carriers against H22 tumor. *Proceedings of SPIE 9324*, 8 pp.

Weiss, B., Davidkova, G., & Zhou, L. W. (1999). Antisense RNA gene therapy for studying and modulating biological processes. *Cellular and Molecular Life Sciences*, *55*(3), 334–358.

Westphal, M., Ylä-Herttuala, S., Martin, J., et al. (2013). Adenovirus-mediated gene therapy with sitimagene ceradenovec followed by intravenous ganciclovir for patients with operable high-grade glioma (ASPECT): A randomised, open-label, phase 3 trial. *Lancet Oncology*, *14*(9), 823–833.

Whelan, J. (2005). First clinical data on RNAi. *Drug Discovery Today*, *10*(15), 1014–1015.

Whitehead, K. A., Langer, R., & Anderson, D. G. (2009). Knocking down barriers: Advances in siRNA delivery. *Nature Reviews Drug Discovery*, *8*(2), 129–138.

Witwer, K. W., Buzas, E. I., Bemis, L. T., et al. (2013) Standardization of sample collection, isolation and analysis methods in extracellular vesicle research: An ISEV position paper. *Journal of Extracellular Vesicles*, *2*. doi: 10.3402/jev.v2i0.20360

Xu, S., Olenyuk, B. Z., Okamoto, C. T., et al. (2013). Targeting receptor-mediated endocytotic pathways with nanoparticles: Rationale and advances. *Advanced Drug Delivery Reviews*, *65*(1), 121–138.

Xu, C., & Wang, J. (2015). Delivery systems for siRNA drug development in cancer therapy. *Asian Journal of Pharmaceutical Sciences*, *10*(1), 1–12.

Yang, S., Che, S. P. Y., Kurywchak, P., et al. (2017). Detection of mutant KRAS and TP53 DNA in circulating exosomes from healthy individuals and patients with pancreatic cancer. *Cancer Biology & Therapy*, *18*(3), 158–165.

Yang, T., Martin, P., Fogarty, B., et al. (2015). Exosome delivered anticancer drugs across the blood–brain barrier for brain cancer therapy in *Danio rerio*. *Pharmaceutical Research*, *32*(6), 2003–2014.

Yang, F., Song, L., Wang, H., et al. (2015). Quercetin in prostate cancer: Chemotherapeutic and chemopreventive effects, mechanisms and clinical application potential (review). *Oncology Reports*, *33*(6), 2659–2668.

Yu, B., Tai, H. C., Xue, W., et al. (2010). Receptor-targeted nanocarriers for therapeutic delivery to cancer. *Molecular Membrane Biology*, *27*(7), 286–298.

Yu, K. H., Zhang, C., Berry, G. J., et al. (2016) Predicting non-small cell lung cancer prognosis by fully automated microscopic pathology image features. *Nature Communications*,*7*, 12474.

Zhong, J., Wen, L., Yang, S., et al. (2015). Imaging-guided high-efficient photoacoustic tumor therapy with targeting gold nanorods. *Nanomedicine*, *11*(6), 1499–1509.

Index

Note: Locators in *italics* represent figures and **bold** indicate tables in the text.

C

D

E

F

G

H

N

O

P

Q

R

S

Printed in the United States
by Baker & Taylor Publisher Services